Britain from space
An atlas of Landsat images

Britain from space
An atlas of Landsat images

R. K. Bullard
R. W. Dixon-Gough

Taylor & Francis
London and Philadelphia
1985

Taylor & Francis Ltd, 4 John St, London WC1N 2ET, UK

Taylor & Francis Inc., 242 Cherry St, Philadelphia,
PA 19106-1906, USA

British Library Cataloguing in Publication Data

Bullard, R. K.
   Britain from space: an atlas of Landsat images.
   1. Aerial photography in geography
   2. Great Britain–Photographs from space

   I. Title. II. Dixon-Gough, R. W.
   551.4′0941   QB632

   ISBN 0-85066-277-X

Library of Congress Cataloging in Publication Data

Bullard, Richard.
   Britain from space.

   Includes index.
   1. Great Britain—Photomaps. 2. Earth—Photographs
   from space. 3. Great Britain—physical geography—Maps.
   4. Landsat satellites. I. Dixon-Gough, Robert. II. Title. III.
   Title: Atlas of Landsat images.
   G1812.21.A4B8 1984   912′.41   84-675252
   ISBN 0-85066-277-X

*Front cover: The natural colour composite mosaic of
Britain produced by the National Remote Sensing Centre,
Farnborough from which the 32 1:500,000 atlas images
have been derived*

*Frontispiece: The 'infrared' version of the composite
mosaic*

*Typeset by Camden Typesetters, London NW1.
Printed in Great Britain by W. S. Cowell Ltd, Ipswich.*

Foreword

*By The Rt. Hon. Lord Shackleton, K.G.,
P.C., O.B.E.*

As Chairman of a recent House of Lords Select Committee Inquiry* into Remote Sensing and Digital Mapping, I am delighted at the production of this new atlas. It was very apparent to my Committee that this whole field was of great importance and that, at the same time, very few people appreciated the significance of the issues involved. The atlas, therefore, is designed to expound and demonstrate the technology of remote sensing and, also, to provide some indication of possible applications of this activity. The range and potential value of these applications is striking indeed. We have scarcely touched the surface, if one can use this metaphor, in regard to space and the digitizing of the information which is already being gained from satellites.

I am glad to note that the Introduction, which deals with remote sensing and details of the Landsat satellite, gives some information on how the images for Britain were obtained, and also provides a useful basic text.

The production of the 32 images covering Britain, together with the text and the specially prepared maps, shows the mapping capability of the Landsat satellite. To help the reader's understanding of the subject the eight specialist images provide further examples of the technology and applications, as do the five examples of other sensors and satellites. Details of both the National Remote Sensing Centre and the Ordnance Survey provide an overview of the services that these complementary government organisations undertake. The section on Geographic information systems shows ways in which data will eventually be grouped together. It will, I believe, be in the synthesizing and co-ordinating roles of geographic information systems that we shall see some of the most important developments in the application of information derived through remote sensing using digital techniques.

The conclusions indicate both the future, with regard to satellites and their improved resolution, and the new applications that can be envisaged. The glossary provides a valuable aid to those unfamiliar with this technology.

The atlas will undoubtedly be of direct value to the teaching profession, but it will also be a medium for stimulating the interest and understanding of the general public.

Shackleton
House of Lords
October 1984

*H.M.S.O., House of Lords Select Committee on Science & Technology, Session 1983-84, 1st Report.

Acknowledgements

This Atlas was brought about with the advice of the Education and Training Working Group (WG5) of the National Remote Sensing Centre. With the opinions of the Working Group and the fullest support of the National Remote Sensing Centre and the publishers, Taylor and Francis, this Atlas has been produced in this format. The Manager of the Centre, Mr Graham Davison and members of his staff are to be thanked for their assistance, particularly Mr Michael Gordon and Miss Tracy Purkiss. Mr Michael Dawes, Publishing Director, Mr David Grist and Mr David Courtney of Taylor & Francis are to be thanked for undertaking the project. Thanks are also due to Mr Keith Eley, Mr Patrick Sorrell and Mr David Wagstaff for their invaluable help.

Dedication

To Pauline and Jennifer, our wives who have helped with this Atlas and encouraged us in its production.

Contents

Introduction

Remote sensing is the detection and recording of information of an object with a sensor without the object and the sensor coming into physical contact using a range of 'recording' devices, which includes the camera. These sensors operate in a wide range of the electromagnetic spectrum (EMS) where the greatest transmission of energy takes place through the atmosphere. These are referred to as 'windows'.

Data collection and storage

The platforms or heights at which the information can be collected with the aid of a sensor can vary from hand-held cameras to sensors aboard satellites thousands of kilometres from the Earth. The examples contained here in this atlas are of images obtained from Earth-orbiting satellites, with the exception of the simulated imagery where the sensor is on board an aircraft.

Figure 1 shows the electromagnetic spectral transmission of energy through the Earth's atmosphere, indicating the percentage of transmission for the spectral regions of wavelength from 0.3 micrometres (μm) to 80 centimetres (cm). The satellites together with their sensors that are described in this publication are shown in Figure 1, the range in which they 'sense' is shown in block form beneath the two sections of the transmission curve.

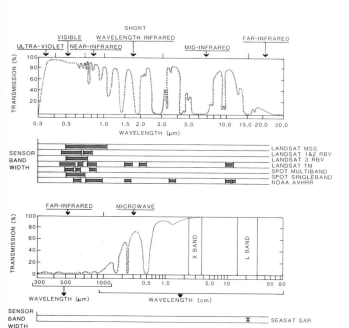

Figure 1. Electromagnetic spectral transmission through the Earth's atmosphere with named spectral regions and the band widths of satellite sensors.
After *Manual of Remote Sensing*, 2nd edn (American Society of Photogrammetry, Falls Church, Va).

The information collected by the satellite-borne sensors has to be transmitted back to Earth. It is therefore necessary that the data is in the form of signals relayed to and received by ground receiving stations. The data collected by a television camera, the return beam vidicon (RBV) camera, for example, has to be converted into numerical form for transmission back to Earth. An object on the ground can be represented by a number which is a value on a grey scale and is determined by the object's reflectance of an energy source, for example, sunlight. Objects on the ground are recorded by the sensors as an area, the size of each area dependent upon the resolution of the sensor. The area is referred to as a 'pixel' or picture element. Figure 2 shows the way that data is collected: the Earth's surface can be considered as a number of coloured building blocks, their size dependent on the resolution of the sensor on board the satellite. The 'full' information of a ground area should include the values which are: a density value; the two-dimensional coordinates (eastings and northings); and the third dimension or height above a datum (usually expressed as height above sea level). The existing satellites do not provide direct means to determine height values but starting with the French Satellite SPOT, relative heights can be obtained, as is explained on page 106.

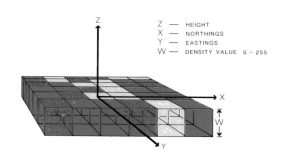

Figure 2. Pixels.

Remote sensing is primarily concerned with the collection of density values and these are obtained by the sensor which allocates a value to the 'colour' of the pixel. The grey scale value, as recorded by the sensor, is depicted in Figure 3. A ground area, Figure 3(a), is represented by a number of columns and lines of pixels where each pixel is represented as a numerical value. In Figure 3(b) the 'spacing' of the columns and lines will be the resolution of the sensor, for example, with a ground resolution of 56 metres by 79 metres for the Landsat multispectral scanner (MSS) the line spacing will be 79 metres while the column spacing will be 56 metres. The numbers allocated to the 'blocks' are the reflectance or grey values. For this example the scale on Figure 3(c) is from 0 to 255 which represents black to white with 256 grey values. The grey scale is shown to be divided into 5 in Figure 3(c) while in fact it can be stretched into 256 levels.

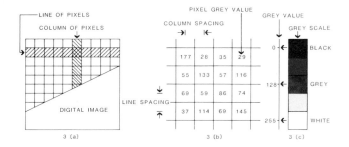

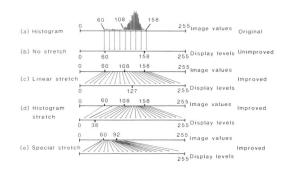

Figure 3. Recording *w* (density) values of pixels.

Figure 5. The principle of contrast-stretch enhancement.

Image analysis

The amount of data that is collected from the sensors is in such large quantities that it is necessary to store the information on magnetic tape. The data can be converted into a photographic format by changing numbers on the tape into their values on the grey scale 0 to 255. Figure 4 shows the conversion of digital data into 'photographic' form. To simplify the process only two numbers are given, 0 and 1 giving a binary representation of the actual object, where 0 represents black and 1 represents white. In Figure 4(*a*) a face has been 'scanned' by a sensor, and in Figure 4(*b*) the face has been converted into 'photographic' form. In practice the image would be in grey tones rather than black and white. The 'photograph', Figure 4(*b*), is unfamiliar as each pixel is shown and the sides of the pixels are straight lines while a face is a series of curved lines. By reducing the size of the 'face' and rounding the edges the 'photograph' can be seen more clearly; this step can also be achieved by viewing the 'photograph' through a lens and then putting it out of focus (softening the edges) or by looking through nearly closed eyelids.

Figure 6. The Bristol Channel (*a*) before and (*b*) after contrast-stretching.

contrast-stretching with Figure 6(*a*) the image as it is received from the sensor and Figure 6(*b*) after the mathematical process of increasing the range. The improvement in contrast is clearly visible.

In Figure 1 the satellites that are referred to are shown to have sensors which are often multi-spectral. The term multi-spectral means that the sensor is designed to record data over more than one region of the electromagnetic spectrum. These regions are shown in Figure 1 below the transmission line. The relationship between the position of the region and the maximum transmission can be seen clearly; these regions are often called windows. Multi-spectral images provide more information for the surface area than with single regions enabling the data to be used for an increased number of applications.

The multi-regions can also be used to produce colour images, the most common being the infra-red composite image. This image may be produced by projecting three regions (Bands 4, 5 and 7) successively on to colour film through blue, green and red filters respectively. Similarly a simulated 'natural colour' image may be produced using the same bands but by using the filter sequence of blue, red and green. Figure 7 shows the combination of three regions to produce colour images: Figure 7(*a*) the infra-red image and Figure 7(*b*) the 'natural colour' image.

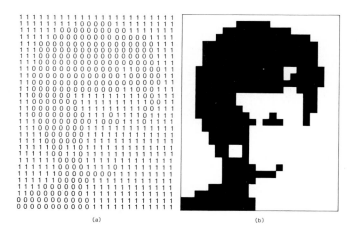

Figure 4. Conversion of binary data into photographic form.

As stated previously the information that is obtained from the sensor does not cover the full range 'black to white'. It is possible by a technique called 'contrast-stretching' to mathematically change the values that are obtained to those that are displayed. Contrast-stretching increases the range that is displayed. Figure 5 shows the limited range in the input (data collected by the sensor) stretched to the output (the 'photograph'). The effect of this process will be to enhance the contrast which makes further analysis of the 'photograph' easier. Figure 6 shows

Applications of remote sensing

It can be said that the applications are only limited by the resolution of the sensor, the availability of the imagery, the techniques used in image analysis, the equipment available for processing the image (hardware and software) and the knowledge, skill and imagination of the user.

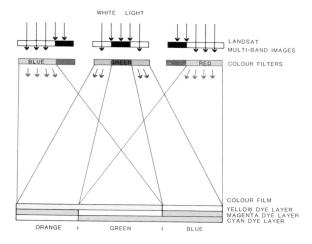

Figure 7(a). Production of an 'infrared' image on colour film.

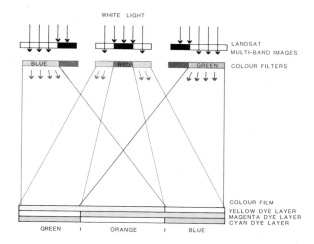

Figure 7(b). Production of a 'natural colour' image on colour film.

With the analysis of the 32 Landsat images of Britain and their accompanying maps, a number of applications are explained. The eight specialist Landsat images provide further examples of the applications of remote sensing as do the five examples of other satellite imagery.

Geology

Because of the area covered by a single satellite image one of the most important applications is that for determining geological information. By being able to compare features over large land surface areas on one image the discrimination of lineaments, faults, dykes and shear zones, for example, become possible where they would not be so obvious on the larger scale aerial photographs.

Land use

Land use mapping, especially the boundaries between urban and rural areas, can be obtained with the aid of satellite. It has been usual in the first generation of satellite images to limit the number of land use classes that are interpreted. With an improvement in the resolution and of the imagery the number of spectral regions (for example Landsat TM has seven), the number of classes can be justifiably increased.

Vegetation

The mapping of the presence of vegetation, its spread and its retreat, can be recorded. Vegetation can be monitored from the repeatable images, weather permitting, and the changes may be indicators of certain factors external to the plants themselves. Drought or disease will cause the vegetation to retreat, as will the damage caused by locusts. To track the path taken by locusts in certain countries their route will be the corridor of devastation they make through the vegetation.

Meteorology

One of the most well-known applications, although not directly related to the land surface, is weather forecasting. The images that accompany the weather forecasts are satellite images, an example of one of these is given on page 99, which shows a NOAA AVHRR weather satellite image.

Pollution

With world-wide concern for saving energy and for the problems of pollution, the monitoring of these can be achieved by Earth-orbiting satellites and their sensors. The 'window' in Figure 1 referred to as the mid-infra-red region is the thermal band where heat differences can be detected. Loss of heat is equated with both loss of energy and pollution. The presence of oil slicks on the surface of the sea combine an example of heat difference (oil and water have different reflective properties) and a pollutant which often kills fish and other marine life.

Bathymetry

The properties of sea water, the presence of pollutants, and in clear water the depth (from 1 to 15 metres), dependent on the reflectance properties of the sea bed and the altitude of the sun, can be determined by interpreting the imagery. The sea lanes of the world can be monitored, coral reefs can be detected as they grow closer to the surface, and submerged wrecks that also present a danger to shipping can be recorded.

Mapping

The boundary between the land and the sea can be clearly defined in the near infra-red bands because of reflective properties of land and water. Satellite imagery provides a small scale mapping facility, not only recording previously unmapped features but correcting those that have been incorrectly recorded. Islands in the Pacific Ocean have been found to be 15 kilometres from their correct position and a major reef 8 kilometres long was 'discovered' with the aid of satellite imagery.

The Landsat multi-spectral scanner

The first Landsat satellite was launched into orbit on 23 July 1972 and since that date there has been a nearly continuous coverage provided by this and subsequent satellites in the series, the first generation, 1, 2 and 3 and the the second generation 4 and 5.

The Landsat MSS images included in this Atlas have been acquired from the first generation satellites with the one exception of the Landsat Thematic Mapper image on page 105.

Figure 8 shows a sketch of a Landsat satellite (Landsat 1 and 2) on which is depicted the main components including the multi-spectral scanner with four bands and the return beam vidicon camera with the three vidicon cameras visible. Solar energy was used to power the satellite and onboard data storage facilities existed which consisted of two wide band video tape recorders (WBVTR) each with a recording capacity of 30 minutes. It takes approximately 27 seconds to record a four-band Landsat MSS scene. This recording facility enabled data to be collected over areas where there was no provision for the information to be transmitted directly back to Earth. As Figure 9 shows the areas covered by ground receiving stations for Landsat MSS images covered almost the total land mass of the world with the major exceptions of East Africa, New Zealand and parts of Asia.

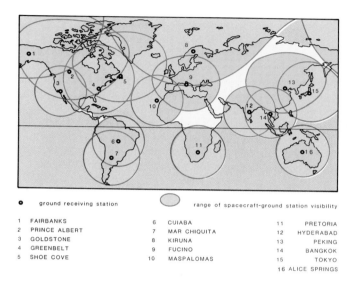

1	FAIRBANKS	6	CUIABA	11	PRETORIA
2	PRINCE ALBERT	7	MAR CHIQUITA	12	HYDERABAD
3	GOLDSTONE	8	KIRUNA	13	PEKING
4	GREENBELT	9	FUCINO	14	BANGKOK
5	SHOE COVE	10	MASPALOMAS	15	TOKYO
				16	ALICE SPRINGS

● ground receiving station ⬭ range of spacecraft-ground station visibility

Figure 9. Landsat ground stations.

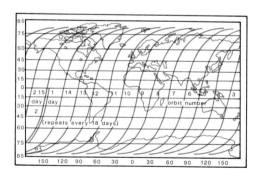

Figure 10. The Landsat ground trace, descending mode.

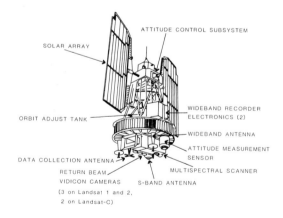

Figure 8. Diagram of Landsat 1/2.

The Landsat satellites 1 to 3 were designed to operate at an altitude of approximately 915 kilometres and take 103 minutes to circle the Earth. At this altitude, due to gravitational forces, the Landsat satellite would be expected to remain in orbit for hundreds of years (though not operational) while a satellite at 250 kilometres would only remain in orbit for 12 days. (It is only at an altitude of 10 000 kilometres that a satellite would remain indefinitely in orbit.) The Landsat scenes cover ground areas of 185 by 185 kilometres. As can be seen in Figure 10 the satellite covers 14 orbits per day covering the Earth's surface from 81°N to 81°S in 18 days, which has led to the expression "round the world in 18 days". The orbit is circular with an inclination to the equator of 99°. This inclination and the time taken to scan an image with the MSS results in a parallelogram-shaped ground area being covered. The sidelap (the area that is common to two adjoining strips) is approximately 14% at the equator, 34% at 40° latitude and 57% at 60° latitude. Each image centre falls within 30 kilometres of the next and in one year the repetition coverage is accurate to within 37 kilometres.

The satellites are Sun-synchronous and cross the equator in the descending mode (going southwards) at approximately 09.30 hrs local sun time (LST) which means that the light conditions (weather permitting), with sun illumination remain the same for adjoining images and also for annual changes. In the ascending mode (going northwards) the satellite crosses the equator at approximately 22.00 hrs when it is dark, so that images will only be collected in the descending mode. With Landsat 3 the thermal band could be used in the ascending mode. For any one point on the Earth's surface (between 81°N and 81°S) the illumination will be the same for each season of the year compared to another year's cycle, thus showing the monitoring capabilities of the satellite's data. The sensors for Bands 4, 5 and 6 are able to record 128 brightness values (0 to 127) while the sensor for Band 7 records 64 brightness values.

Landsat 2 was put into orbit to provide a nine-day interval between the two satellites (1 and 2). Although Landsat 1 was only designed for a life of one year it was finally 'retired' on 6 January 1978, after nearly 5½ years, or 5½ times the life expectancy. Landsat 2 was launched on 22 January 1975 and operated for seven years. Landsat 3 was put into orbit on 3 March 1978 and is still operational.

The spectral sensitivity of the Landsat MSS bands is shown in Figure 1 and in more detail in Figure 11. Each of the four bands has six detectors which means that six strips on the Earth's surface are scanned at the same time. The scene contains 2340 scan lines representing a total of 390 scans taking place for each image. Each line contains 3240 pixels which means that each band contains approximately 7·5 million pieces of information and each four-band scene contains approximately 30 million.

As well as obtaining the position of the satellite in space, ground control is also required to rectify the image. For many of the distortions shown in Figure 13 they require ground data in the form of x and y co-ordinates to make the corrections. Unlike an aerial photograph which needs a minimum of four control points per stereoscopic overlap, positioned as shown in Figure 13(a), in the UK the satellite images require approximately 110 ground control points per scene, as shown in Figure 13(b). Although a certain amount of sidelap (dependent on the latitude) occurs, as stated above, the Landsat MSS images have not been used to obtain heights (z), but forthcoming satellites like SPOT, see page 106, have stereoscopic capability and will therefore be able to be used for heighting.

The 32 images of the United Kingdom in this book are taken from a mosaic produced by the National Remote Sensing Centre. The pixels in the mosaic are 100 metres. The mosaic was obtained from many Landsat 1 and 2 scenes recorded over a number of years, which were 'joined together' to eliminate cloud and mathematically 'smoothed' to eliminate joins. This mosaic was used in preference to individual images, which might have had better resolution, but would not provide continuity throughout the atlas for the 32 images.

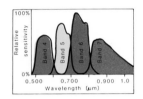

Figure 11. The spectral sensitivity of Landsat MSS bands.

The scanner, because it does not record the image like a camera in one single exposure, is prone to movement while it is scanning, not least because of the oscillation of the scan mirror. A number of corrections need to be applied to the scanned images. Figure 12 shows the Landsat MSS geometric distortions that occur due to movement of the satellite relative to the Earth as the image is being scanned. The position of the satellite has to be carefully tracked and its position sufficiently well known to point the ground receiving antenna towards the satellite with an accuracy of better than one degree of arc. The parabolic antenna of the ground station will need to be at least 12 metres in diameter to receive the data from the satellite. The data are collected directly when the satellite is in line of sight and over 5° above the horizon of the ground receiving station.

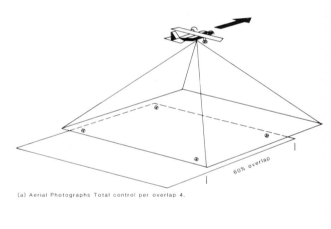

(a) Aerial Photographs Total control per overlap 4.

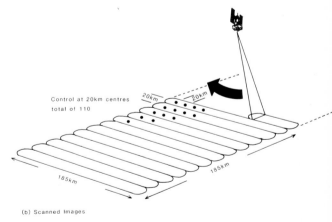

(b) Scanned Images

Figure 13. Ground control to rectify images.

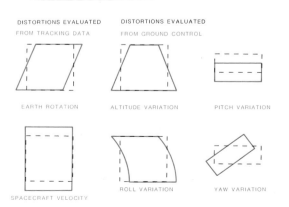

Figure 12. Landsat MSS geometric distortions.
After *Manual of Remote Sensing*, 2nd edn (American Society of Photogrammetry, Falls Church, Va).

About the Atlas

The greatest advantage of the images contained within this Atlas is that they generally contain far more detail, more accurately than a conventional map of a comparable scale, particularly details relating to land cover. Atlas maps of a similar scale are usually very generalized, simplified impressions of the Earth's surface. Satellite imagery, subject to its graphic resolution, will contain all the information that can be sensed on or near to the Earth's surface and therefore presents an almost overwhelming amount of data. Because of the novelty of producing an Atlas based upon satellite imagery, it was decided to include, opposite the image, an interpretive map and a short section of descriptive text.

The maps used in this Atlas have been compiled directly from the satellite imagery and have been drawn to a common specification. Each map is orientated in the same direction as the image and has been drawn at half the scale to allow for the inclusion of text. All maps and images are orientated with the northern edge at the top of the page unless indicated otherwise.

The detail included on the maps is principally of a topographic nature and any evidence of soil types, underlying geological trends and vegetation cover, other than woodland, present on the image, is dealt with in the text. Because the map detail has been derived directly, only

that detail recognizable on the satellite image is included, subject to the rules of generalization imposed by the reduced size of the map image. When the detail is reduced, it is impossible to portray it on a 1:1 ratio and consequently it has to be simplified, symbolized and, in the case of adjacent areas, occasionally combined. Readers may notice that small areas of woodland have been omitted whilst other areas in close proximity to one another have been combined. Similarly it may have been necessary to artificially widen some long thin lochs to allow the light blue infill to be included, whilst linear features, such as the coastline, have been simplified. There was also the problem of defining the coastline, particularly in marginal regions, such as Essex where it is difficult to draw a line between land and sea in the lower lying areas. Generally speaking, the dividing line was taken to be the limits of vegetation, mud or silt being included within the sea.

The most important function of the maps is to enable the readers to identify the patterns, shapes, colours and lines that appear on the satellite images, and it is therefore essential that they relate to the image. Whilst there is always a temptation to include extra detail such as National Parks and the full drainage system, it was felt that the maps should be kept as simple as possible. If physical

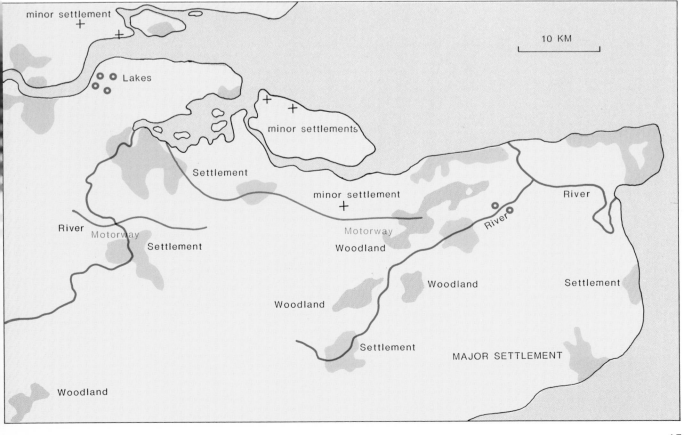

features such as Dartmoor and the North Downs can be clearly identified on the images, their extent and position on the map is indicated by their names.

Many names have been abbreviated. The basic convention adhered to relates to the size of settlement. These settlements that have an areal extent are written in full whilst smaller settlements, having no recognizable extent, are depicted with a cross and a form of abbreviation. A similar convention is used for hydrographic features.

Whilst maps have been included within the Atlas, the reader is encouraged to make full use of Ordnance Survey maps when examining the images. In particular the 1:250,000 Routemaster Series will be found very useful Those Ordnance Survey Routemaster maps relating to each image are referred to and their location is included on the index map.

In keeping with the simplified nature of the maps symbolization has been kept to a minimum, as illustrated below, and is only used when easily identified features on the image are too small to allow their geographical extent to be depicted. Alongside the maps accompanying the five specialized images, keys are included to facilitate the interpretation of those maps.

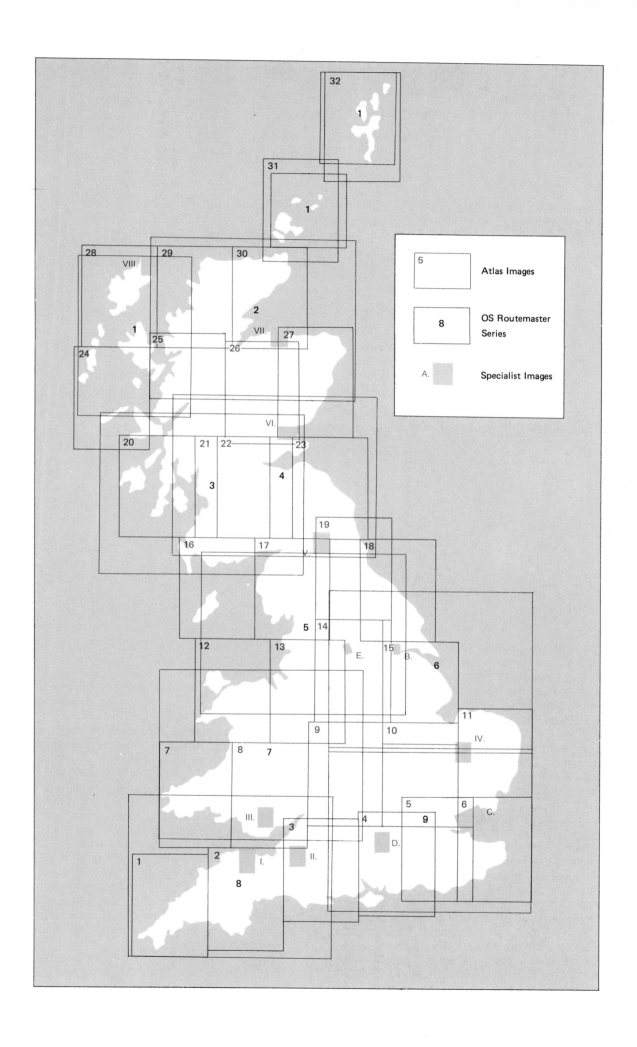

5 Atlas Images

8 OS Routemaster Series

A. Specialist Images

1. Cornwall and the Scillies

For reference see O.S. Routemaster Series, Sheet 8

Bo   Boscastle
Fo   Fowey
He   Helston
Li   Liskeard
Lo   Lostwithiel
Mu   Mullion
Pd   Padstow
Pe   Penryn
Po   Porthleven
Rd   Redruth
Wa   Wadebridge

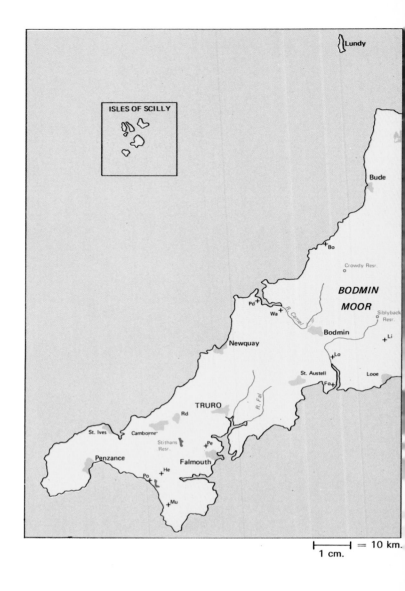

In the accompanying Landsat MSS image, the remote nature of the extreme south-west peninsula is superbly illustrated. One factor that becomes very clear is the lack of arterial communication links caused by the series of granitic domes (the lighter areas on the opposite page) down the centre of the peninsula. This, together with the poor agricultural conditions found in the large areas of moorland, required inhabitants to seek sustenance from the sea. The final factor that influenced the development of this area is the old, resistant mineral-bearing rocks which led to important mining industries, particularly copper and tin.

The most obvious features that can be recognized are the high granite domes of Bodmin Moor and Henbarrow Downs. The very light colour of these two areas is due to the thin acidic soil covering the impermeable granite which results in a moorland habitat with sparse vegetation cover. These areas are incapable of supporting any farming activity other than very rough grazing and is the major reason for the lack of settlements in these areas. An additional feature found in Henbarrow Downs is the presence of china clay or kaolin. This has been formed from the chemical weathering of felspar crystals in the granite and results in a greyish-white substance. This is extensively mined and processed in the St Austell area and these activities will result in a lack of vegetation cover and the corresponding white image.

Evidence of another mining area may be found to the

north of Camborne and Redruth. Here, a slightly lighter area than the surrounding green is probably the spoil heaps from the famous Cornish tin mines. Cornwall was once the chief supplier of tin until the end of the 19th century when cheaper imported ores made the deep Cornish tin mines uneconomic to work.

Broadly speaking, the remaining area can be divided into two zones. The brownish-green areas consist of fine loamy soils overlying Devonian sandstone deposits. These areas are suited to dairy farming and stock rearing with crops such as potatoes and broccoli near to the coast. The lighter green areas, particularly around Penzance, are fertile, loamy soils overlying granite. Unlike the soils overlying the granite of Bodmin Moor, in this area the abundant felspar has improved the soil.

The mainland peninsula has also helped shape the character of the Scilly Isles. It may be observed from the Landsat MSS image that several of the islands appear to be joined. This is in fact shallow water which links many of the islands at low tide, caused by debris eroded from the mainland being washed out to sea and deposited around the islands.

A well known coastal feature found around this peninsula is the presence of rias or drowned river valleys caused by a lowering of the land relative to the sea level. Several of these can be seen on this particular image, characterized by the winding, branched creeks. These include the estuaries of the Fal, Fowey and Helford rivers.

2. Devon

For reference see O.S. Routemaster Series, Sheet 8

As	Ashburton
BR	Burrator Reservoir
BS	Budleigh Salterton
Bu	Buckfastleigh
CM	Combe Martin
CR	Clatworthy Reservoir
Cr	Crediton
Da	Dawlish
Dm	Dartmouth
FwR	Fernwood Reservoir
GT	Great Torrington
Hn	Honiton
Ho	Holsworthy
Il	Ilfracombe
Iv	Ivybridge
Kb	Kingsbridge
MR	Meldon Reservoir
No	Northam
Ok	Okehampton
Pa	Paignton
Pl	Plympton
Pm	Plymstock
To	Torquay
Tp	Torpoint
Sa	Saltash
We	Wellington

├———————┤ = 10 km.
1 cm.

This scene covers one of the most popular tourist areas of the British Isles, with the upland area of Dartmoor sandwiched between the coastal resorts of north and south Devon. Also visible in the centre top is the picturesque Exmoor National Park.

Perhaps the most dominant area of this scene is the igneous granitic dome of Dartmoor, the highest area of southern England. The lighter portions of the Landsat MSS image correspond to the thin acidic soils of the moorland area, sparsely covered with vegetation and offering poor grazing and pasture. Soils are very thin except in the river valleys where evidence of better soils, together with more sheltered growing conditions, may be deduced from the darker colouring representing denser vegetation and tree cover.

To the north of Dartmoor the area of darker green roughly corresponds to the carboniferous deposits of very resistant millstone grit. Here the better quality, loamy soils are more suited to farming, particularly permanent grassland, stock rearing and dairying, together with some cereal production.

The fertile soils of south Devon, to the east of the Exe valley and around Barnstaple are indicated in bright green on the Landsat MSS scene.

The Devonian sandstones of Exmoor produce another area of high moorland, generally unfavourable to agriculture and therefore containing few settlements of any size. Most farms in this area, as with those around Dartmoor, are located in the sheltered, more productive valleys where the breeding of store cattle and sheep farming is very important.

Examination of the coastline reveals several interesting features such as the drowned river valleys of south-west Devon and large areas of sand in north Devon. Drowned river valleys, and their winding tidal creeks, may be found at Plymouth (the Tamar), Kingsbridge and Dartmouth. The large areas of sand and sand dunes found at Braunton Burrows and Morte Bay (to the north of Barnstaple) are at right angles to the dominant wave direction and prevailing winds, whilst the headlands separating these bays indicate the east-west strike of the Devonian rocks.

Most of the larger towns are located near to the coast. Plymouth, the largest town and the regional centre of the area, has an excellent natural harbour but unfortunately, due to its isolation from the main industrial areas of the British Isles, has failed to become an important commercial port. Strategically, however, it is an important Royal Naval Base guarding the Western Approaches of the English Channel. The second most important town is Exeter, situated at the lowest bridging point of the River Exe. This is a nodal point for most routes passing into the South-West Peninsula and is also the principal market for east Devon. Along both coastlines are many seaside resorts, of which the largest is Torbay. It is sheltered from northerly winds, lying in the lee of Dartmoor, and has a relatively low rainfall. Resorts along the north Devon coast are smaller than their counterparts in south Devon principally due to their isolation from major routes, which during the nineteenth century tended to lead towards south Devon.

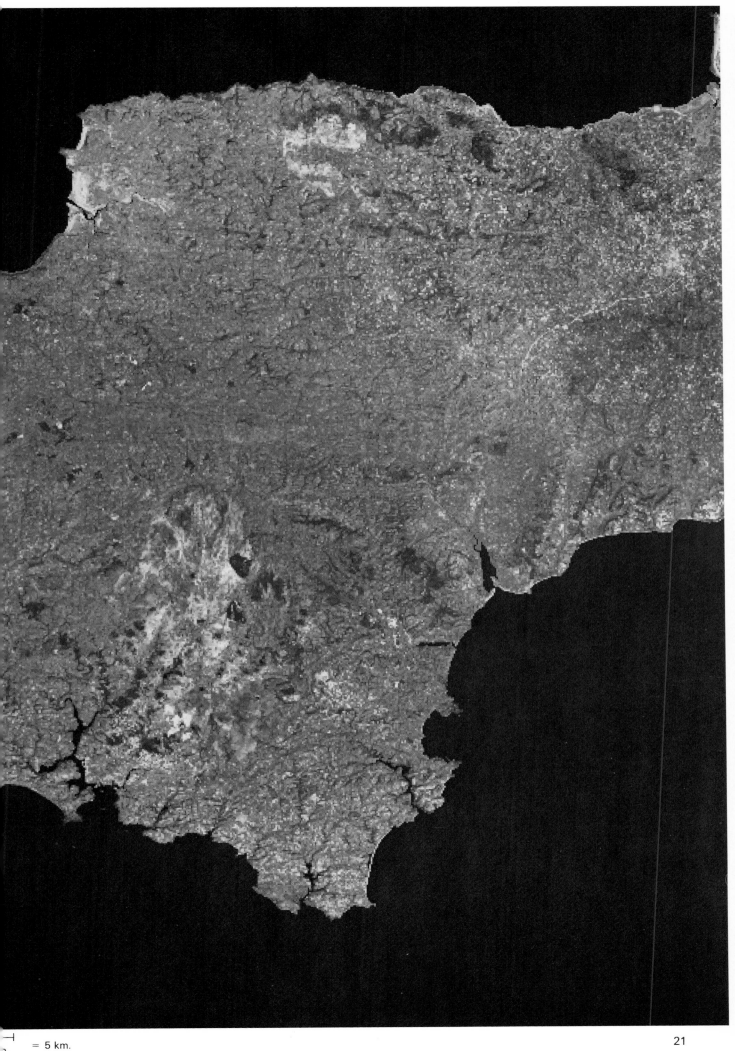

3. The New Forest to the Severn Estuary

For reference see O.S. Routemaster Series, Sheets 8 and 9

BF   Blandford Forum
BL   Blagdon Lake
BoA   Bradford on Avon
Br   Bridport
Ca   Calne
Ch   Chepstowe
Co   Corsham
Cr   Crewkerne
CR   Cheddar Reservoir
CVL   Chew Valley Reservoir
De   Devizes
Fe   Ferndown
Gl   Glastonbury
Il   Ilminster
Ke   Keynsham
LR   Lyme Regis
LSW   Llanwern Steel Works
Ma   Marlborough
Me   Melksham
Na   Nailsea
Nl   Nailsworth
Po   Poole
Ra   Radstock
Ri   Ringwood
Sh   Shaftesbury
St   Street
Sw   Swanage
Te   Tetbury
Th   Thornbury
Ve   Vernwood
Wa   Wareham
WB   Wootton Basset
We   Westbury
WM   Wimborne Minster

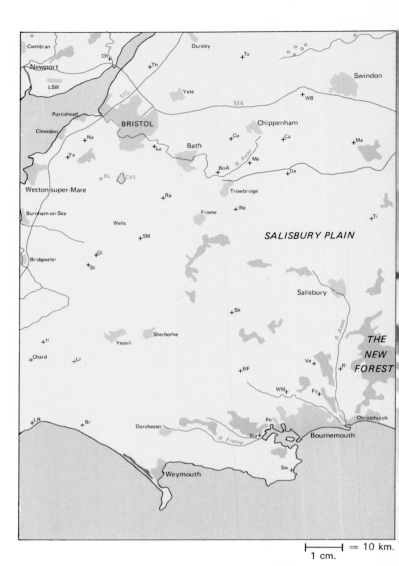

├───┤ = 10 km.
1 cm.

The Landsat image may be divided into five geological zones. In the south-east are the tertiary deposits of the Hampshire and Dorset Basin, the brown area on the opposite page representing the Dorset heathland and the New Forest where the soil coverage is thin and infertile. This area can be further divided by the fertile valley of the River Avon. To the east is the New Forest where the gravel-topped plateau supports moorland, bracken and hardy grasses. This area is about 200 square kilometres and consists of approximately 40% wooded areas, 58% open heath and 2% cultivated land. To the west of the Dorset River Avon is the generally lower lying heathland whose soil characteristics are inferior even to that of the New Forest. Surrounding this zone is a lighter green area which represents a band of London clay, offering more fertile soil and improved farming conditions.

The second major zone is the cretaceous deposits of the chalklands forming Salisbury Plain and the Downlands around Marlborough and to the west of Swindon. This can be seen as the brown-green area starting to the north of Weymouth and continuing in an arc towards the eastern edge of the image. Here the most dominant crop is cereal production, with some dairying and sheep farming with extensive areas given over to military use.

The third major zone is that of the Jurassic rocks running in an approximately north/south direction from Bridport in the south, towards Tetbury in the north. These form the Blackmoor Vale and eventually the Cotswolds. The soils covering this area are largely well drained and loamy, suited to cereal and root crop production together with dairying.

To the west is the fertile vale of Somerset, seen in a slightly brighter green. Here the soils are particularly rich and deep and market gardening is gradually replacing dairying. Just to the north is the bright green area representing a flat, low-lying peat bog over alluvial deposits known as Sedgemoor. Much of this area has now been drained and is used as pasture land for beef fattening around Bridgwater and dairying around Cheddar. The Somerset Plain is interrupted to the north by the Mendip Hills, composed largely of carboniferous limestone, rising to a height of about 300 m. The darker green of the Mendip Hills contrasts strongly against the brighter green of the Somerset Plain: soil coverage is very thin, which, when coupled with the bleak aspect, makes this area more suited to cattle and sheep rather than arable farming. Just to the north of Cheddar and Cheddar Reservoir can be seen two white areas representing large limestone quarries.

The major settlements can be clearly located on the image. Along the south coast are the seaside resorts of Bournemouth and Christchurch and the ports of Poole and Weymouth. To the north, the most important settlement is Bristol, for over 500 years the second most important port to London due to its monopoly of trade to the west. Despite its decline as a port, it remains the second largest city in southern England.

Other interesting features that can be clearly seen on this image are the sand banks in the Severn Estuary and the shingle beach of Chesil Bank. This extends 30 km in a gentle arc from West Bay to the Isle of Portland, enclosing for part of its length the Fleet, a brackish lagoon.

4. Central Southern England

For reference see O.S. Routemaster Series, Sheet 9

Am	Amersham
B	Beenham
Bb	Bembridge
Be	Beaconsfield
Bt	Barnet
Ca	Camberley and Frimley
Ch	Chesham
Eg	Egham and Staines
Ep	Epsom
F	Farnborough
Fa	Farnham
Fr	Fareham
Fw	Fawley
GC	Gerrards Cross
Ha	Hatfield
Hd	Hampstead
Hl	Hayling Island
Ho	Hook
HoT	Henley-on-Thames
Hu	Hungerford
Hv	Havant
Hy	Hythe
Hz	Hazlemere
Le	Leatherhead
Lm	Lymington
Lp	Liphook
Ma	Marlow
Np	Newport
Ov	Overton
PB	Potters Bar
Rc	Richmond
Ri	Rickmansworth
Ry	Ryde
Sb	Stubbington
Sh	Shanklin
Su	Sutton
Th	Thame
Ux	Uxbridge
Wa	Watford

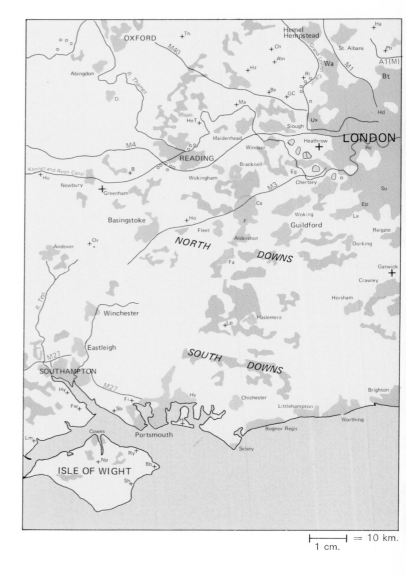

The most dominant features on this scene are the Hampshire and London Basins separated by the chalk Downlands. The Hampshire Basin, located in a broad arc around the Solent stretching from Portsmouth to the western edge of the image, is much darker than the adjacent chalklands. The New Forest, to the west of the Solent, can be clearly picked out as an area of brown and dark green.

Enclosing the Hampshire Basin are the South Downs, stretching from just to the north of Havant and extending eastwards towards Brighton. Particularly noticeable are the wooded areas on the north-facing scarpface of the South Downs. To the north may be seen the lighter area of the North Downs with its south-facing scarp slope extending from Basingstoke to the eastern edge of the image.

The London Basin is enclosed between the dip slopes of the North Downs and the Chilterns. This major feature may be clearly seen on the adjacent image as a darker area sandwiched between the lighter North Downs and darker tree-covered Chilterns. It is covered by Tertiary deposits of London clay and areas of sand together with extensive Quarternary deposits of gravels and silts. Evidence of the gravel deposits may be indicated by the number of water-filled gravel pits in the river valleys around Reading and Rickmansworth.

The northern limits of the chalk can be seen as a slightly darker colour running in a north-east/south-west direction, from the centre top of the image down to the centre of the western edge. Unlike the Downs, the Chilterns have extensive tree cover occupying the superficial deposits of moist, flinty clay.

In addition to the principal physical features that may be interpreted from this image, a wealth of other detail can be very clearly seen. For example, the entire scene, with the exception of the Downlands, is very extensively wooded. The two main areas are those around Haslemere and Bagshot. The soils around Haslemere are very acid, sandy, well drained and unsuitable for most agricultural purposes. Similar soils occur also around Bagshot, hence the extensive tree cover.

Communication systems are well defined and motorways can be picked out clearly. Some, for example the M27, are more clearly defined than others. The probable reason for this anomaly lies in the radiance of the surface material, the M27 being laid in concrete whilst the M3, for example, is surfaced using a tarmacadamed material, thereby offering less contrast with surrounding detail. Other features consisting of a high proportion of concrete are also clearly visible. Heathrow and Gatwick airports can be easily spotted and the strategically important sites of Greenham, Aldermaston and Burghfield are also evident.

With respect to the drainage of this region, the valleys of the Rivers Test and Itchen, for example, may be picked out because of their contrasting colour. With patience, the River Thames can be traced from London to beyond Oxford, whilst the reservoirs by Heathrow Airport are very obvious.

5. London and the South Coast

For reference see O.S. Routemaster Series, Sheet 9

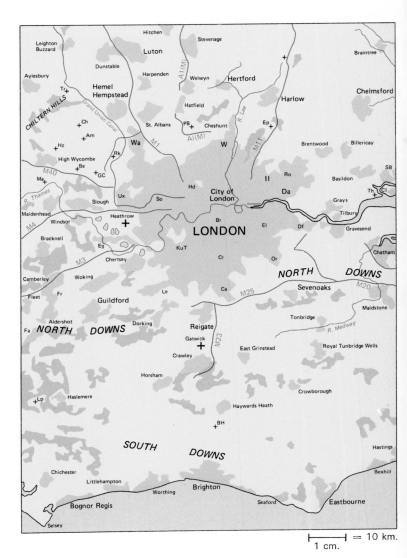

1 cm. ⊢——————⊣ = 10 km.

This scene is an excellent section across southern England, from the Chiltern Hills to the chalk of the South Downs. Between these outcrops may be seen the syncline of the London Basin and the denuded dome of the Weald. Each topological feature can be seen on the image as a distinctive band of colour, whilst in the centre is London with its sprawling suburbs.

The north-west corner of the image is occupied by the light green area of the Vale of Aylesbury. This is backed by the darker coloured Chiltern Hills, the northern outcrop of chalk that dips beneath the London Basin to reappear as the North Downs. The dip slopes of the Chilterns are overlain with deposits of flinty clay which support an extensive cover of hardwood trees. These can be seen as dark green patches and examples may be observed to the north of High Wycombe. The darker area to the south-east of the Chilterns is the London Basin bounded to the south by the lighter band of the North Downs. West of London, the London clay is covered by deposits of alluvial gravel and evidence of water-filled gravel excavations may be seen. Along the western edge of the image are the dark green areas of heath and woodland around Bagshot, Camberley and Liphook.

In the southern portion of the image, the areas that comprise the Weald can be picked out easily. The dark green area extending from Bexhill westwards to Horsham and then eastwards to Royal Tunbridge Wells is the High Weald. Surrounding this is a lighter band representing the clays of the Low Weald, whilst to the south, the band running from just south of Liphook towards Eastbourne is the chalk of the South Downs. This and the North Downs form the rim of the Weald.

Cutting across the centre of the image is the River Thames, the largest and deepest estuary on the south-east coast, flowing from central southern England. The Thames has very high tides with a rise of 4–6 m as far up river as London Bridge. The combination of high tides and a deep estuary means that ocean-going ships can sail 65 km into the heart of the country. The wide tidal range necessitated the construction of docks in the cheap, readily available marsh-land bordering the river. Some of these docks may be seen on the image, the long Royal Victoria and Royal Albert Docks and West India Dock in the Isle of Dogs.

London lies in the hub of the area depicted in this scene. The built-up area almost fills a circle 50 km in diameter whilst the London region extends nearly 160 km across. Much of the London region lies below 60 m, but the low-lying ground either side of the River Thames is interrupted by numerous gravel topped hills which became the early growth points for London during Roman times.

Within the London region it is possible to divide the area into a number of land use and settlement zones. For example, the commercial centre of London may be identified, together with extensive areas of parkland, the docks and the industrialized Lee Valley. To the north-east, Epping Forest and to the west, Heathrow Airport are easily identified.

= 5 km.

6. East Sussex, Kent and Essex

For reference see O.S. Routemaster Series, Sheet 9

AR   Abberton Reservoir
Ba   Battle
Bl   Brightlingsea
BoC Burnham-on-Crouch
Br   Broadstairs
Bt   Brentwood
Ch   Chatham and Rochester
Fa   Faversham
Gr   Grays
Ha   Hawkwell
HR   Hanningford Reservoir
Hy   Hythe
Ki   Kingsdown
Ly   Lydd
Ma   Maldon
Mi   Minster
NF   North Foreland
NR   New Romney
Ro   Rochester
Ry   Rye
Sa   Sandwich
SB   South Benfleet
Sh   Sheerness
SlH Stanford le Hope
Te   Tenterden
Th   Thameshaven
Wa   Walton-on-the-Naze
Wh   Whitstable
Wi   Wickford
WM   West Mersea

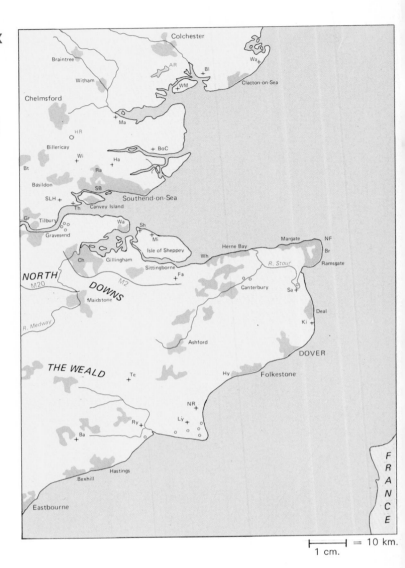

├────────┤ = 10 km.
1 cm.

This image is an excellent illustration of how geology can influence the nature and characteristics of a region. Contrasts can be made between the shape of the coastline of Essex and the Thames Estuary with that of Kent and East Sussex. Similarly the image may be classified according to colour, and a number of distinctive zones are easily recognized. If a geological map of this area is examined, there is a close correlation between the shapes of the coastlines, the areas of colour and the underlying geological structure.

The most logical and distinctive division between zones is the Thames Estuary. To the south of the Thames are the North Downs and the Weald, whilst to the north is the relatively low lying area of Essex, with its complicated coastline of winding creeks and mudflats. The area to the south of a line drawn approximately between Gravesend and Sandwich was once the western end of a low dome of chalk, extending across the Straits of Dover into France. The highest parts of the dome were eroded to expose the underlying rock structure of clay and sandstone. Subsequently, the denuded dome was breached by the sea, cutting Britain off from the Continent. The permeable chalk and harder sandstone were resistant to the forces of erosion and formed lines of hills whilst the clay belts were worn down into broad vales.

The dark area, seen towards the bottom left of the image, is the resistant sandstone of the Weald. During Roman and Saxon times the area was covered with forest and substantial clearance of the area was not undertaken until Elizabethan times, when much of the forest was reduced to charcoal to smelt the local iron ores or used to make ships. To the east of the Weald is the lighter area of Romney Marsh extending into the sea, which is mainly the result of consistent additions of shingle ridges, mostly of flint, with alluvium deposited on them. Bordering the Weald, curving from Romney Marsh towards Maidstone, is the lighter coloured Vale of Kent, a mixture of pasture and arable. To the north is a darker area running in a band from just south of the River Thames and widening towards the coast. This is the chalk of the North Downs, an area once noted for sheep rearing.

With respect to the coastline, the harder, more resistant rocks of Kent and East Sussex produce a smoother shoreline. Chalk is very resistant to erosion by the sea and this may be observed where the North Downs meet the sea, between Folkestone and Sandwich.

In contrast, the Essex coastline consists of low-lying flooded shores. This area on the image is extensively covered with London clay. As a result the rivers occupy broad, shallow valleys. This coast is slowly sinking, causing the sea to spread up the valleys to form a very winding, complicated coastline, with salt marshes that are covered at high tide. Examples of these may be seen as the light brown areas around West Mersea.

7. South-west Wales

For reference see O.S. Routemaster Series, Sheet 7

Aa   Aberaeron
Ca   Cardigan
Na   Narberth
NQ New Quay
OR Oil Refinery
Pe   Pembroke
Sa   Saundersfoot

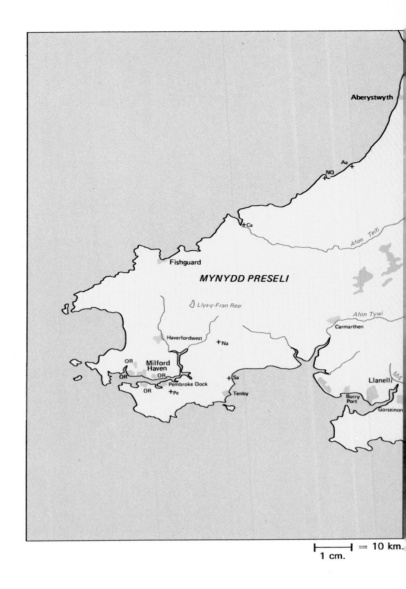

⊢———⊣ = 10 km.
1 cm.

This large peninsula in many ways epitomizes the topology and characteristics of Wales, with an upland core, surrounded by low coastal plateau surfaces. These lower regions, which are more extensive to the south, are broken by the Rivers Teifi and Tywi and also by the extensive drowned valley of the Cleddau at Milford Haven. The region has a long history of human activity dating back to Neolithic times and was a major passage-way used by traders and missionaries, travelling eastwards as long ago as the Bronze Age. During the immediate post-Roman era, when missionaries of the Celtic Church were very active in areas bordering the Irish Sea, many monastic cells were established, the most famous being St David's. Most of the area is sparsely populated, the principal industrial areas being in the south, centred around Milford Haven and Llanelli.

The dull green area east of Fishguard represents the higher areas of Mynydd Preseli and Mynydd Pencarreg, rising to a height of between 330 and 520 m. The rocks used to construct Stonehenge were transported from this area. The darker green areas are the extensive woodlands found to the north-east of Carmarthen, whilst the brown areas represent peat.

The band of bright green running east of Carmarthen is the Vale of Tywi, considered to be one of the most important dairying regions of Wales. Although it initially supplied the industrial areas of South Wales with butter and milk, improved transportation has led to much milk being supplied to as far afield as London.

Agriculture is important throughout the peninsula. Along the coastal parts, where rainfall is relatively low, arable farming is important with much of the interior given over to grassland. In the south, the more favourable climate has influenced the development of market gardening. The light soils warm up very quickly, encouraging the growth of early potatoes and broccoli.

The most important development in this area has been the evolution of Milford Haven into Britain's major oil port. The numerous creeks and tributaries of the drowned River Cleddau make this the largest area of sheltered water in the British Isles. A new dredging scheme has enabled this port to handle 250,000 ton oil tankers, on all tides, at the oil terminals, which can be seen on the image.

In the south-east of the image, sandwiched between the Gower Peninsula and the extensive areas of sand at Pendine and Pembrey, are the industrial complexes of Burry Port and Llanelli. These are set on the western edge of the South Wales coalfield and are important for their metallurgical industries, particularly tin plating. The coastal strip of this area is very sandy and these deposits are clearly visible. The estuaries of the Tywi and Loughor are both heavily silted, whilst the sands of Rhossili Bay on the Gower Peninsula are the result of wave action and the prevailing winds. To the south of Rhossili Bay can be seen the resistant headland of carboniferous limestone and Worm's Head.

⊢ = 5 km.

31

8. Mid and South Wales

For reference see O.S. Routemaster Series, Sheet 7

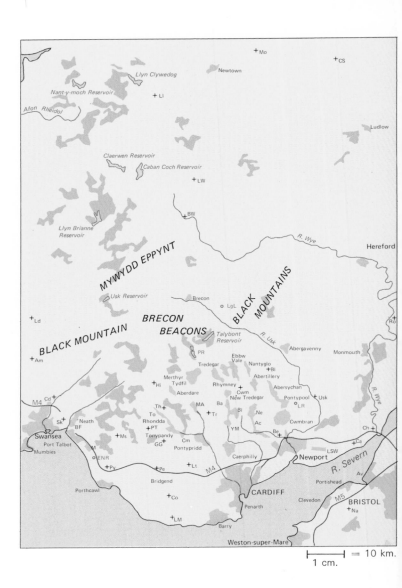

⊢———⊣ = 10 km.
1 cm.

Wales is a country of contrasts and these are evident in this scene. To the south is the industrialized South Wales coalfield with its associated industries, whilst immediately to the north are the barren, inhospitable Brecon Beacons with their harsh climatic conditions. The area north of the coalfield is often referred to as the Heartland of Wales, and consists of the bleak Cambrian Mountains. These gradually slope downwards to the east to form the Borderlands or Welsh Marches.

The image forms a striking picture of the barren areas of the Heartland, a region of high relief, generally exceeding 210 m in height. The soils are acid and shallow and large areas of hill peat exist, especially where the soils are developed on impervious substratum such as boulder clay. These can be seen as the large brown areas on the Cambrian and Black Mountains. The high rainfall of the region, combined with steep slopes and regular run-off, provide ideal conditions for water storage, both for domestic and industrial purposes and flood control. Evidence of numerous, large man-made reservoirs can be seen. Forestry plays an important part in this region and extensive areas of dark green conifer plantations can be seen in the Brecon Beacons and along the Cambrian Mountains.

The eastern part of the image, still structurally part of the Welsh massif although it includes parts of the English border counties, may be seen as a lighter area on the image. It is a transitional zone with no definite north/south boundaries and tends to merge gently into the English lowlands. The light green area around Hereford is the Hereford Basin extending westwards towards the edge of the South Wales coalfield and the upland masses of the Black Mountains.

Further south towards the estuary of the River Severn may be seen the dark green of the heavily wooded Wye Valley and Forest of Dean.

Towards the south coast of Wales the coalfield can be easily defined, sandwiched between the light green Vale of Glamorgan and the light brown Brecon Beacons. The South Wales coalfield is dissected by deep parallel river valleys running in a NNW/SSE and ENE/WSW direction, which have to a large extent been determined by faulting. The development of coalmining has been restricted to these valleys, causing considerable crowding and linear settlement patterns with hardly any break between neighbouring villages. The proximity of the coalfields to the sea led to an important export market, the docks of Barry, Cardiff and Swansea being built largely for this purpose. Ironstone, associated with Lower Coal Series, in the north and north-east of the coalfield, led to the development of the Welsh iron and steel industry. The industry was originally located adjacent to the deposits of ironstone and coal, but obsolete blast furnaces and the need to import iron ore has led to its transfer to the coastal areas.

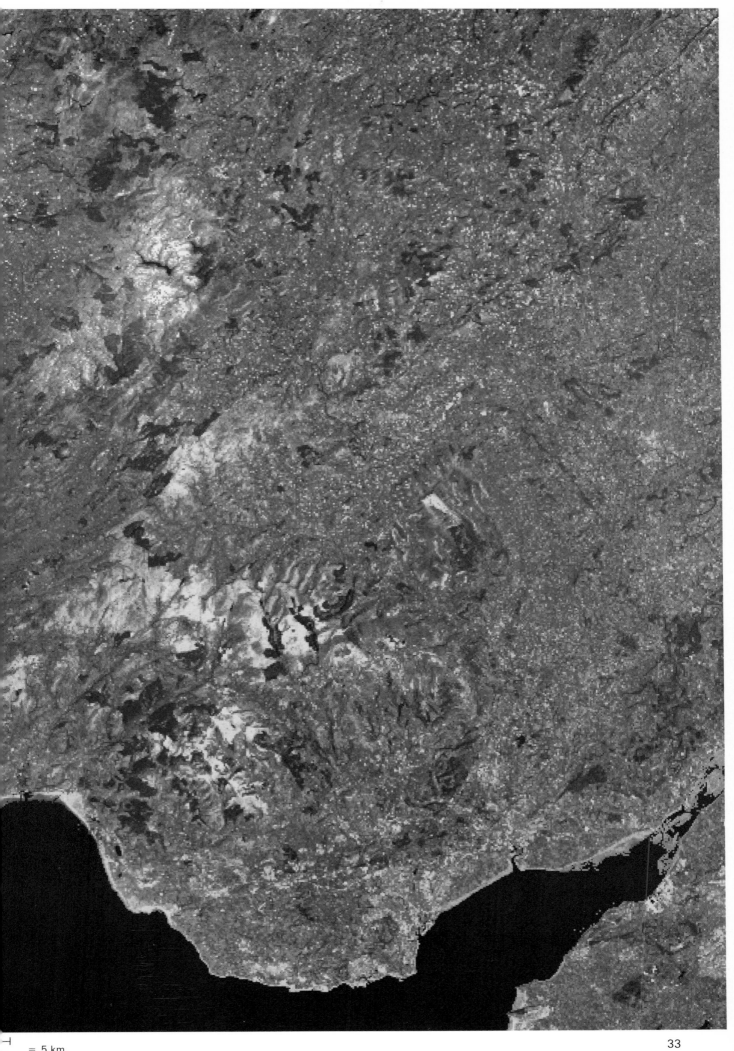

9. Birmingham and the West Midlands

For reference see O.S. Routemaster Series, Sheets 6, 7 and 9

As	Ashby-De-La-Zouch
At	Atherstone
Be	Bedworth
Bi	Bicester
Bn	Bridgnorth
Br	Brackley
BR	Blithfield Reservoir
Ca	Carterton
Ch	Chasewater
Ci	Cinderford
CN	Chipping Norton
Co	Coalville
Do	Dorridge
Du	Dudley
DW	Draycote Water
Ha	Halesowen
Ki	Kidlington
Kn	Knowle
Le	Letterworth
Ly	Lydney
Na	Nailsworth
Ne	Newport
Pe	Pershore
Ro	Ross-on-Wye
SC	Sutton Coldfield
Sm	Smethwick
So	Solihull
S	Stourbridge
St	Stroud
Te	Tetbury
Th	Thornbury
Wa	Walsall
WB	West Bromwich
Wi	Witney
Wo	Wolverhampton

This image depicts the relatively low lying Western Section of English Midlands. The Midland Plain is represented by a roughly triangular section bounded to the west by the older, more resistant rocks of the Welsh Borderland and to the south and east by the scarplands of the Cotswolds. Structurally, the higher ground to the west of the Midland Plain can be described as part of the Welsh massif, even though substantial portions of this scene, for example the south Shropshire highlands and the hills of Herefordshire, are politically not part of Wales. This area can be clearly seen on the Landsat MSS image as a gentle arc of dark green, ranging from the darker Forest of Dean in the south towards Telford in the north of the image. Similarly the Cotswold Hills can also be easily defined as a dark green band running from the bottom edge of the image, around Tetbury, in a north-easterly direction towards Banbury and Daventry. The north-west-facing scarpface can be clearly seen against the light green areas of the fertile lower Severn Valley and the Vale of Evesham. The south-east of the Cotswolds is the brown area representing the Vale of Oxford.

The lighter coloured Midland Plain is dominated by the light brown coloured mass of the West Midlands conurbation, situated towards the centre top of the image. Surrounding this urban area are the river valleys of the Severn from the western edge of the plateau, the Avon joining the Severn at Tewkesbury and the Trent which flows northward towards the East Coast. It is interesting to note the difference between the Trent, which occupies a wide, light green valley, and the Severn and Avon, with narrow valleys, particularly in their upper reaches.

The West Midlands conurbation of Birmingham and its two large suburban areas of Solihull and Sutton Coldfield, together with the other units which comprise the Black Country, covers an area of about 700 square kilometres with a population of approximately 2·5 million people.

Birmingham is one of the few major trading centres, and one of the largest cities in the world, not situated on a navigable river. It was a small, unimportant market town that owes its growth to the Industrial Revolution and its close proximity to a rich seam of coal which outcropped over wide areas ranging from Dudley to Stourbridge, iron ore which existed in the coal measures and limestone in the Dudley area. To facilitate the transportation of the industrial products, a web of canals were built linking Birmingham with the remainder of the country. This area has remained in the forefront of transportation developments since that time. The railways followed, to a large extent superseding the canals, whilst with the introduction of the motorways Birmingham plays a vital role as a hub to the system. These motorways can be seen on the image, some being more obvious than others. The M69 linking Coventry with the M1 near Leicester is particularly noticeable owing to construction work taking place at the time when the image was sensed.

10. London and the Fens

For reference see O.S. Routemaster Series, Sheets 6 and 9

Am   Amersham
Ap   Ampthill
Bi   Biggleswade
Bo   Bourne
BS   Bishop's Stortford
Bu   Buckingham
C   Chesham
Ch   Chatteris
Ep   Epping
Ho   Holbeach
Hz   Hazlemere
Li   Littleport
MD   Market Deeping
NP   Newport Pagnell
Oa   Oakham
PB   Potters Bar
PR   Pitsford Reservoir
R   Ramsey
Ra   Raunds
Ri   Rickmansworth
Ro   Romford
Sa   Sandy
SW   Saffron Walden
Wa   Watford

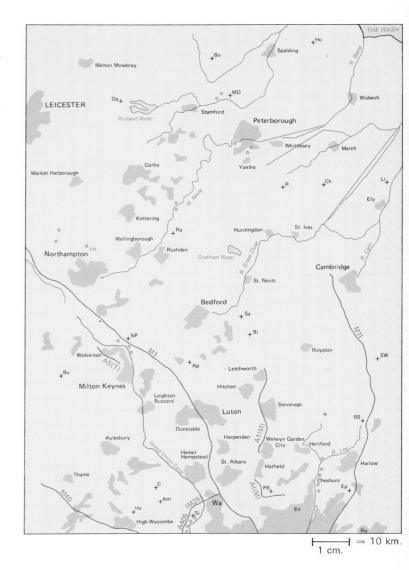

This area can be divided into two major zones. In the north-east is the dark green area of the Fens, through which several large rivers flow en route to the Wash, just visible in the extreme north-eastern corner of the image. To the south are the northern suburbs of London and the commuter belt, which effectively stretches as far north as Bedford. The boundary of the Fenlands is very pronounced. They are enclosed to the west by the Kesteven Plateau and to the south by the Jurassic Northampton Uplands. During pre-glacial times the Wash was probably a broad shallow valley draining the area now occupied by the Fens into the smaller North Sea. Meltwater later caused a rise in sea-level to submerge the Wash and penetrate the Fenland Basin, leaving a large area of marshland. Currents and tides deposited marine silts on the seaward side whilst the rivers, which had maintained their courses across the shallow basin, deposited alluvium. Reed marshes, which flourished behind the coastal silts and between the rivers, were converted into peat fen whilst coastal subsidence led to the formation of silty fen. Some of the underlying Clay Vale was not covered with peat or silt and provides the island sites of March, Ely, Spalding and Wisbech.

Artificial drainage of this area has been carried out since Roman times and the long straight drainage channels can be seen on the image. Major attempts at reclamation came during the 17th century when Dutch engineers straightened out the old river courses. Examples include the New Bedford River, east of St Ives, and the straightened Nene near Peterborough. The reclaimed land is fertile and the climate and soils allow intensive farming. The soils of the peaty fen are alkaline and rich in minerals. There is no true rotation, and the main crops consist of potatoes, sugar-beet and wheat. The silt fen has a greater variety of crops but the emphasis is on fruit farming and market gardening.

To the west of the image are the Northampton Uplands, broken by the upper courses of the Welland and Nene. The two most valuable rocks in this area are the marl stones and Northampton sands. Both rocks yield substantial quantities of iron ore although the latter are more important in output, producing nearly 11 million tons per year. Until comparatively recently, iron ore production had resulted in no large urban concentrations. The production centres being scattered and the workers living in agricultural villages. The initial problem lay with the lack of coal so the ore was sent out of the area. Blast furnaces were eventually set up at Kettering, Wellingborough and Market Harborough to make pig iron and an integrated iron and steel works built at Corby. This has since been closed.

In the centre of the image are the Bedfordshire Lowlands, part of the Inner Clay Vale extending from the Wash to the Bristol Channel. Further to the south-west can be seen the sweep of the Chilterns extending north-east as far as Hitchen.

= 5 km.

11. East Anglia

For reference see O.S. Routemaster Series, Sheet 6 and 9

AR Abberton Reservoir
Be Beccles
Br Brightlingsea
Bu Bungay
Ca Caistor-On-Sea
Cr Cromer
Di Diss
DM Downham Market
Fa Fakenham
Fe Feltwell
Fr Framlington
Go Gorleston-On-Sea
Ha Hadleigh
Hr Harleston
Hu Hunstanton
Hw Halesworthy
Ma Manningtree
Mi Mildenhall
Sh Sheringham
So Southwold
Su Sudbury
Sw Swaffham
Ti Tiptree
Wa Walton-On-The-Naze
We Wells-Next-The-Sea
Wo Woodbridge
WM West Mersea
Wy Wymondham

1 cm. = 10 km.

In East Anglia superficial glacial and other recent deposits obscure the underlying solid rocks. These deposits form a number of well marked regions that can be observed on the image. This is one of the most highly cultivated regions of the British Isles due to the combined factors of the climate and the Norfolk system of crop rotation. Over this region the maximum rainfall occurs during the early summer months, balancing evaporation due to high temperatures and aiding summer crop growth. Winters tend to be very cold and the frosts break up the soils, making preparation easier. Norfolk rotation was introduced during the Middle Ages and refined during the 18th century into a 4-year cycle of seeds, wheat, roots and barley.

The largest single area that can be observed is the bright green sweep of land stretching from the western edge of the image, between Haverhill and Braintree, in a wide northward arc towards East Dereham and Norwich. This is the low, undulating East Anglian plateau of boulder clay intermixed with chalks and sands, divided into two (High Norfolk and High Suffolk) by the Rivers Ouse and Waveney. In places the deposits may be 30–50 m thick. The soils in this area are fertile, light loams. Barley, wheat and sugar-beet are the main crops but cattle and pigs are important, particularly in Suffolk.

To the north of this region is a darker green area extending in an approximate east/west direction to meet a darker, north/south line near to King's Lynn. This is the low-lying area of north-east Norfolk and the Goodsands.

These regions are separated by the Cromer Ridge, running in a south-west direction and the remains of a terminal moraine, consisting of glacial sands and gravels, rising to a maximum height of 90 m. The Goodsands is an area of chalky boulder clay and glacial sands. The soils are light and have been improved by Norfolk rotation, during which clover is ploughed into the soil. Between this area and the sea, marshes exist behind barrier beaches and spits.

Along the north-western edge of the image may be seen the dark green area of the flat, peaty Fens, whilst adjacent to this, running in a north/south direction, are the East Anglian Heights and the Norfolk Edge, separated by the lower Brecklands. The Norfolk Edge forms a low escarpment of lower greensand facing towards the Fens and the Wash. This can be seen as a thin dark green line east of King's Lynn. Around Thetford are the Brecklands, a lower area drained towards the Wash, covered with glacial sands and gravel. This is a very distinctive light area on the image with large green patches representing extensive areas of afforestation. Further south are the bright green areas of the East Anglian Heights, rising to a height of 92 m on Newmarket Heath. These are formed by an outcrop, dipping to the east.

The area lying between Norwich and the sea, crossing the lower courses of the rivers Bure, Waveney and Yare, is the Norfolk Broads. The extensive Broads are formed from old, water-filled medieval peat cuttings that are gradually being encroached by vegetation.

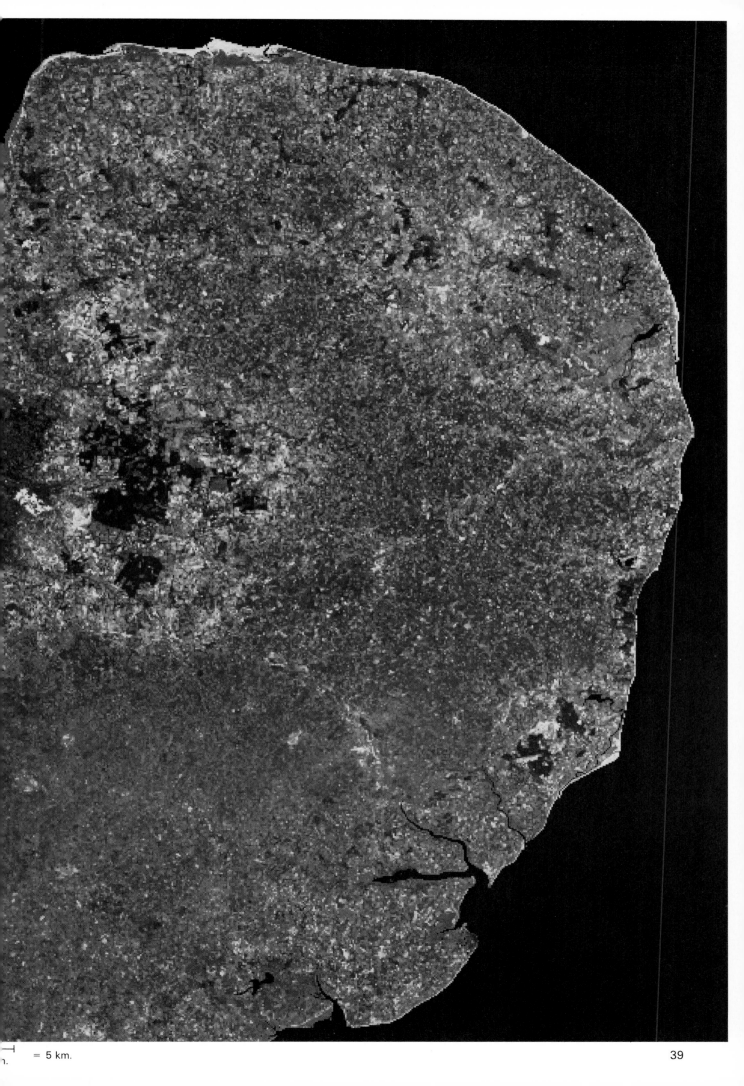

12. North Wales

For reference see O.S. Routemaster Series, Sheet 7

Am   Amlwch
Be   Bethesda
BF   Blaenau Ffestiniog
Co   Conwy
Do   Dolgellau
Lf   Llanfairfechan
LJ   Llandudno Junction
Lr   Llanrwst
MB   Menai Bridge
Pe   Penmaenmawr
Po   Porthmadog
Pw   Pwllheli
Ty   Tywyn

├────┤ = 10 km.
1 cm.

North Wales can be divided into three physical regions: the heartland; the west coast and Anglesey; and the North Wales coast. Slate quarrying is the industry long associated with these areas and at their peak the annual output exceeded 200,000 tons. Evidence of quarries can be seen on the image as lighter areas, particularly around Snowdonia and Blaenau Ffestiniog. Unfortunately, the industry has declined, largely owing to the use of substitute building materials. One of the most important values of this region lies in its high rate of rainfall and physical conditions suited to water storage. Many reservoirs have been built in this area and provide domestic water supplies for areas outside Wales such as Merseyside and the Midlands. Conditions are ideal in this area for the generation of hydro-electricity. Forestry is an important occupation in the heartland and large areas of dark green conifer plantations can be seen on the image.

The west coast region follows the edge of Cardigan Bay and includes the marshy area around the artificial Lake Trawsfynydd. Farming is the major occupation and the settlement pattern has evolved through this. Each of the broad river estuaries and basins had their local farms and coastal trade and hence two types of settlements evolved, the marketing centre and the small port. In the Dwyryd basin the distribution centre is Ffestiniog whilst Porthmadog became the port for this northern region. Similarly, the comparable settlements for the Wnion are Dolgellau and Barmouth.

Anglesey and the North Wales coast is structurally an area of wave-cut platforms occurring at different heights. The main platform, the Menaian, with an average height of 82 m, can be traced through the Lleyn peninsula and Anglesey and on to the coastland as far as Llandudno. In the east, the platform narrows as the Cambrian Mountains come close to the sea. The dominant branch of agriculture throughout the region is pasture, although lighter drift soils may be used for crops. Beef cattle, dairying and sheep are reared and the upland areas are used as rough pasture. Dairy production is particularly important on Anglesey and the Lleyn peninsula. One of the largest milk processing plants is at Llangefni which provides a source of milk supply to the chocolate factory at Bournville. To the east of the Afon Conwy, the area's close proximity to the industrialized Midlands and South Lancashire has helped develop the large tourist resorts of Llandudno, Colwyn Bay, Rhyl and Prestatyn. Elsewhere in the region, the tourist industry is less important and only the railhead of Pwllheli can compare with the resorts of the north coast.

The light areas, seen around Snowdonia, represent clouds, the shadows of which can be seen as the darker zone to the north-west of the cloud.

= 5 km.

13. The North Midlands and the North-west

For reference see O.S. Routemaster Series, Sheets 5, 6 and 7

AE   Alderley Edge
Al   Altrincham
Bo   Bolton
Br   Brymbo
BR   Blithfield Reservoir
Bu   Bury
Cd   Cheadle
Ce   Cefn-Mawr
Ch   Chapel-en-le-Frith
Cl   Chorley
Cr   Crosby
Da   Darwen
EP   Ellesmere Port
Fl   Flint
Gl   Glossop
HG   Hazel Grove
Ho   Holywell
Hl   Hoylake
Ki   Kirkby
Kn   Knutsford
Ll   Llay
MD Market Drayton
Mi   Middlewich
Mo   Mossley
MSC Manchester Ship Canal
Na   Nantwich
Ne   Newcastle-under-Lyme
NM New Mills
Ol   Oldham
Or   Ormskirk
Po   Poynton
Ra   Ramsbottom
Rh   Rhosllanerchrugog
Ro   Rochdale
Sb   Stalybridge
St   Stone
We Welshpool
Wh Whitchurch
Wi   Wilmslow
Wm Wem

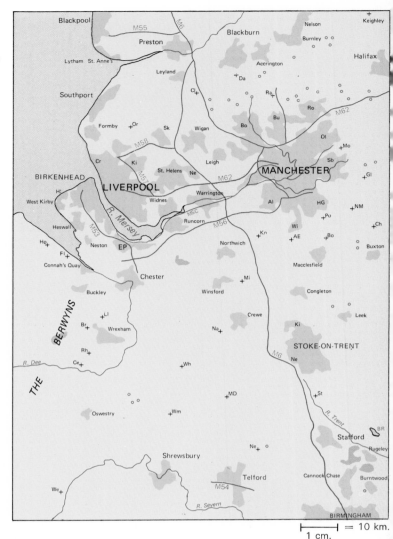

$\vdash\!\!-\!\!-\!\!-\!\!\dashv$ = 10 km.
1 cm.

This is an area of great contrasts, ranging from the industrialized Midlands northwards across rural Shropshire and Cheshire to the extensive Merseyside and Manchester conurbations. The contrast is equally pronounced in an east/west direction, from the Pennines across the flat Cheshire and Shropshire plains to the Berwyn Mountains of Wales. It is possible to identify five major zones. Approximately 50% of the area is coloured light green, occupying a wide sweep of land running from the Black Country northwards to Lancashire, at the top of the image. This represents the relatively low lying, fertile plains of Shropshire, Cheshire and Lancashire and includes both the Wirral and Fylde peninsulas. Cutting across this is an almost continuous area of light brown representing the industrial complexes of Merseyside, north Cheshire, south Lancashire and Manchester. To the west of the image may be seen the dark brown of the barren, higher parts of the Berwyn Mountains with a pocket of industrialization in its foothills, centred around Wrexham in the North Wales coalfield. Similar dark brown areas may be observed along the eastern edge of the image, indicating the acid, peat moors of the Pennines. These curve in a wide crescent around Manchester to separate it from the final zone, the traditional cotton towns of east Lancashire.

It is interesting to compare the adjacent estuaries of the Dee and Mersey. From Roman to Tudor times, Chester was the most important port on the north-west coast, but silting of the broad-mouthed estuary of the River Dee led to its decline. Despite efforts made to revive the port trade by excavating a new channel through the Dee sands during the 18th century, trade was diverted to the expanding port of Liverpool, which grew rapidly, largely because of its easier connections with the Manchester area. The major reason why the Mersey did not suffer a similar fate to that of the Dee lies in the constriction of its channel to a mere 1 km width. This increases the speed of the tidal flow to 7 knots, effectively scouring the bottom of the channel. Liverpool is now Britain's second port.

Manchester, linked to Liverpool by the Manchester Ship Canal, is the natural communication centre for south-east Lancashire. Many of the rivers of this area converge here, and railways were later built to connect the city with its hinterland. Surrounding Manchester are a number of towns built where the rivers leave the moorlands and enter the Lancashire plain. Around these towns occur the two natural products that led to the location of the cotton industry, water and coal, which traditionally provided the power for the cotton mills. Urban sprawl has now effectively combined these towns in the Manchester conurbation. The sprawling growth along the valleys and the lines of communication are easily identified.

The line crossing the image in an east-west direction south of Stoke-on-Trent is caused by a flaw in the data used to produce this image.

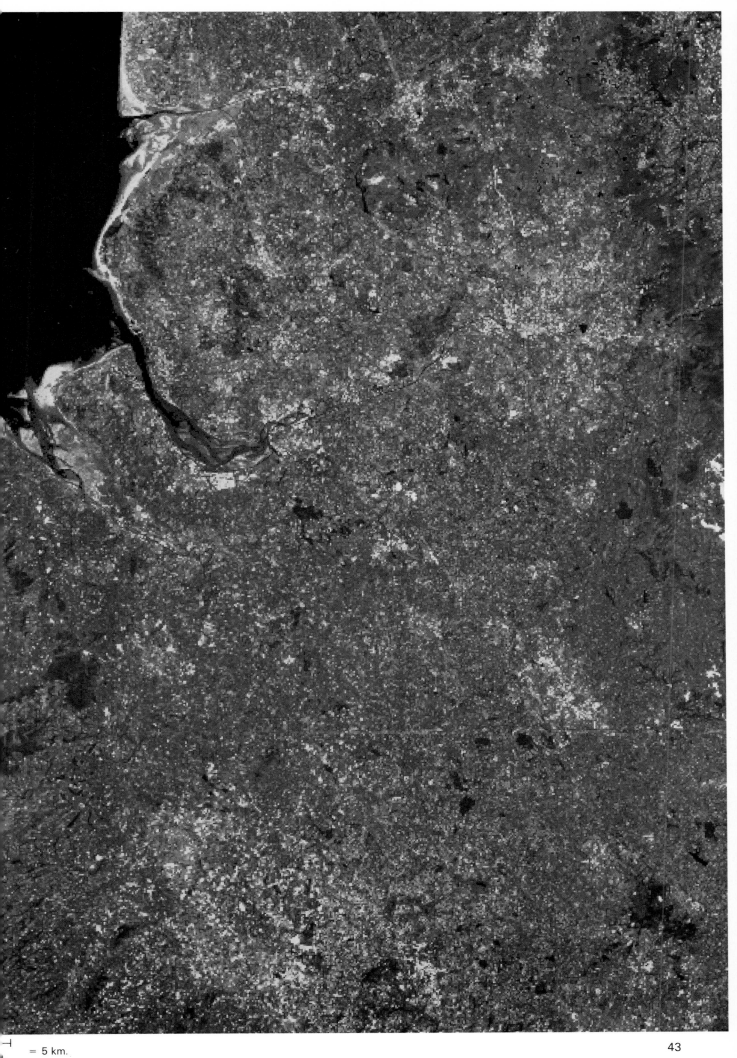

14. The Southern Pennines

For reference see O.S. Routemaster Series, Sheets 5, 6 and 7

AE Alderley Edge
Ba Barnoldswick
Bi Biddulph
Bo Bolton
Bp Belper
Bu Bury
CC Clay Cross
Ch Chapel-en-le-Frith
Cl Clitheroe
Gl Glossop
Ha Haslingden
He Hemsworth
HN Hoylake Nether
Kn Knutsford
Mo Mossley
NM New Mills
No Normanton
Ol Oldham
Ot Otley
Ro Rochdale
Rw Rawtenstall
Sa Salford
Sb Stalybridge
Se Selby
St Stockport
Tc Tadcaster
Ut Uttoxeter
We Wetherby

Running northwards through the centre of this image are the Pennine Uplands, consisting almost entirely of millstone grit and carboniferous limestone. It is possible to differentiate between these two rocks on the image. The dark brown areas represent the very acid peat moorland, found on top of the impermeable millstone grit, whilst the grasslands of the well drained carboniferous limestone shows up as bright green. The boundary between these two rock types is in a line running eastwards from the south of Manchester to Sheffield. Along the western margin of the Pennines lie the coalfields of Staffordshire and Lancashire and to the east, those of Derbyshire, Nottinghamshire and Yorkshire.

Although the Pennines form a distinctive area, devoid of settlements of any size and largely uncultivated, they have many conflicting functions, such as water catchments for domestic and industrial supplies, sheep pastures, grouse moors and recreational areas for the large industrial zones that border them. Plentiful supplies of soft water are available in the millstone grit area and the numerous small reservoirs may be seen in the adjacent image.

The Pennines effectively divide two great industrial regions, Lancashire to the west and Yorkshire to the east, seen as the extensive areas of light brown on the image. Both are based upon coalfields and have textile related industries.

The core of industrial Lancashire is Manchester, which has developed into a sprawling conurbation. Lancashire is essentially the home of the mill-based cotton industry, initially located in the Pennine valleys in the east and south-east of the county. The attraction of this area lay in its plentiful supply of soft water combined with high humidity and a close proximity to sea ports and chemicals for dyeing and bleaching the cloth. This industry has suffered a severe decline since World War II and many towns are attempting to attract alternative industries.

Further south is the industrialized area of North Staffordshire and the Potteries. Based upon the Staffordshire coalfield with its high quality coking coal and readily available clays and marls, the pottery industry has remained in this area. It is an excellent example of localized industry, since over 80% of all pottery workers live here.

On the eastern side of the Pennines are the extensive industrialized areas of West and South Yorkshire and the East Midlands. The woollen industry is traditionally centred in a conurbation stretching from Leeds to Huddersfield and from Bradford to Wakefield. In addition, the area is important for garment manufacturing and engineering. Sheffield and Rotherham are famed as steel manufacturing centres.

The north-east corner of the image is occupied by the light green of the Vale of York, a flat area with a predominantly agricultural population. York, the chief town of northern Britain until the industrial revolution, is situated on an up-raised area of gravel astride the River Ouse.

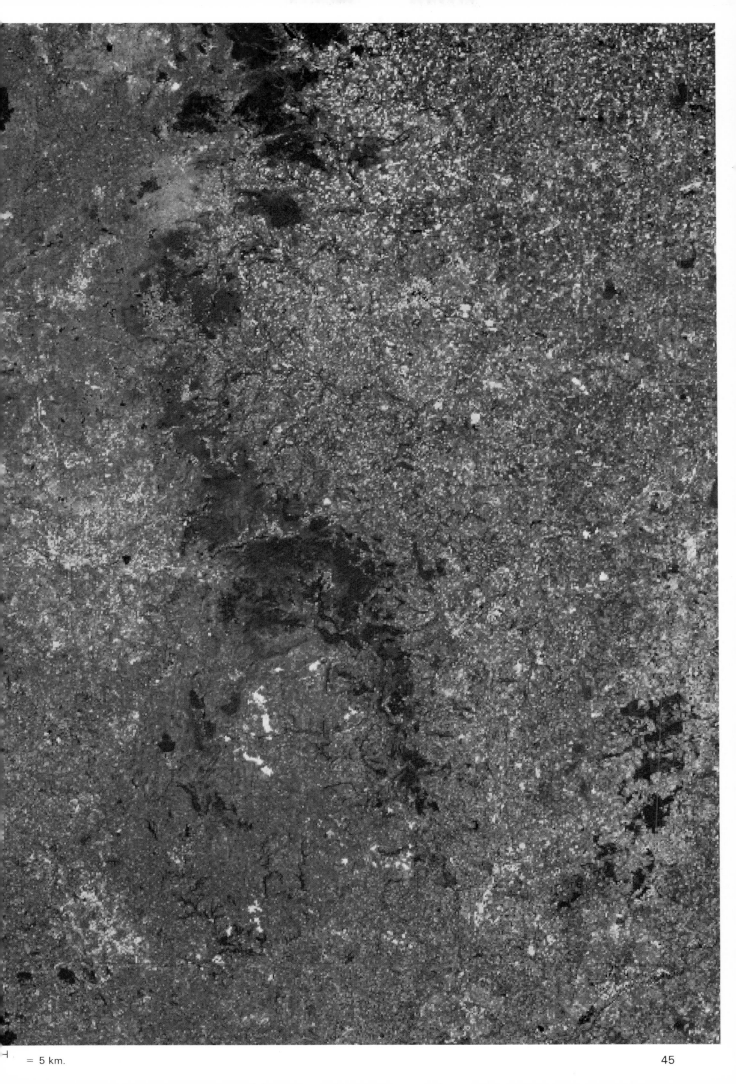

15. Lincolnshire and Humberside

For reference see O.S. Routemaster Series, Sheet 6

Al	Alford
Ba	Barton-upon-Humber
Bg	Brigg
Bo	Bourne
Br	Broughton
Cl	Cleethorpes
CL	Chapel St Leonards
Ha	Harworth
He	Hedon
Hc	Horncastle
Ho	Holbeach
Hu	Humberston
Im	Immingham
Ke	Keyworth
Ma	Mablethorpe
MD	Market Deeping
Mo	Moorends
Mr	Market Rasen
Oa	Oakham
Se	Selby
Sl	Sleaford
Tp	Thorpe
Wa	Waddington
Wi	Withernsea
WS	Woodhall Spa

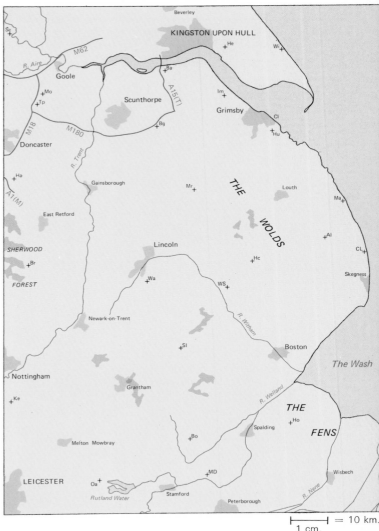

├───┤ = 10 km.
1 cm.

To the north of this image is the Humber estuary, the largest undeveloped estuary in the country and now the site of regional planning. The largest town along the estuary is Kingston-upon-Hull, which had no direct link with South Humberside and the South of England until the opening of the Humber road bridge in 1982. Although not built when the image was recorded, its location may be defined from the position of the A15 north and south of the Humber estuary. The M62, linking Humberside to the Midlands and South Yorkshire, can also be seen on the image. Along the northern shoreline of the estuary are the light coloured sandbanks and long spit of Spurn Head.

To the west of the image, running in a north/south direction, is the darker area of the Vale of Trent. These lowlands are mainly of Keuper marls, overlain with boulder clay giving local variations in relief. They are linked to the Vale of York, seen along the northern edge of the image, by the levels of the Humber Head. This is a very flat area of alluvium with characteristics similar to those of the Fens. To the west of the Vale of Trent may be seen the dark green area of Sherwood Forest, situated on an area of Bunter Sandstone. In the northern part of the Vale, the course of the River Trent cuts into an outcrop of the lower Lias. It is in this region that ironstones have been exploited for more than a century and was a major factor in the development of Lincolnshire's iron and steel industry centred around Scunthorpe. The ore is mined using open cast methods and the two principal mines are at Santon

and Dragonby. These can be seen on the image as the small white areas to the east of Scunthorpe. In the southern part of the Vale of Trent, between Newark and Nottingham, may be seen the Vale of Belvoir.

Running north/south, parallel to the Vale of Trent, is the oolitic limestone area of the Lincoln Heath and the Kesteven Plateau. The Lincoln Heath has no glacial covering and, although once an area of bracken and gorse, is now under cultivation. The Kesteven Plateau, extending eastwards from Leicester towards the Fens, is much broader and higher and has a mantle of boulder clay.

The Lincoln clay vale can be seen in a similar dark green extending from the Humber to the Fens. In the north it is compressed between the Lincolnshire Heathland and Wold, but tapers out south of Lincoln. It is made up of a succession of Jurassic clays, the majority of which are covered with deposits of boulder clay.

Between the clay vale and the sea is the lighter area of the Lincolnshire Wolds. In the northern section there are few glacial deposits and a distinctive west facing scarp face. Further south, the Wolds become broader and the scarp face is dissected by streams. Agriculture in the northern Wolds is restricted owing to difficulties in obtaining water supplies, but the soils are fertile.

Around the Wash are the Fenlands, one of the largest areas of continuous flat land in Britain.

46

16. Galloway and the Isle of Man

For reference see O.S. Routemaster Series, Sheets 3 and 5

Ca   Castletown
Ga   Gatehouse of Fleet
LD   Loch Dornal
LO   Loch Ochiltree
PE   Port Erin

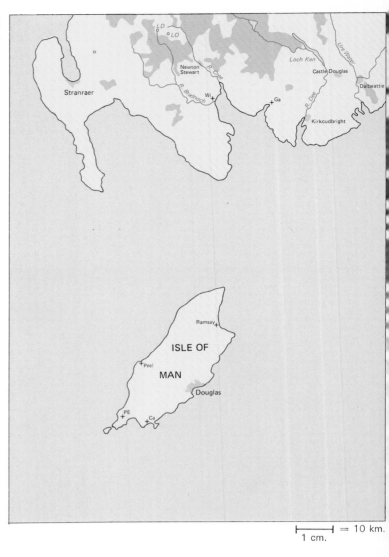

⊢———⊣ = 10 km.
1 cm.

Galloway, hemmed in to the north and east by unbroken moorland, has been largely isolated from the mainstream of Scottish affairs. Its wide tidal inlets led to close cultural links being forged with Ireland and the western islands and peninsulas. Many of the inhabitants of this region originate from Antrim, as do the cattle for which Galloway is famous.

Much of the lower regions have rich land, particularly the Rhins and the lowlands connecting it to the mainland around Stranraer. The region has the 'softer' climate associated with the west coast, a relatively heavy rainfall of 1270 mm and a high humidity. Oats and turnips form the main crop where the drift and alluvial soils permit cultivation, but most of the land is under permanent or rotational grass for Ayrshire or Galloway cattle. Beyond the low-lying areas and the steeper well-drained slopes, peaty soils and wet moorland vegetation predominate. They are used as rough grazings for the hardy Blackface and Cheviot sheep. Since World War II, however, much of these peaty areas have been planted by the Forestry Commission. The high rainfall combined with the acid nature of the peat are admirably suited to the rapid growth of Sitka spruce.

The settlements in this area reflect the once important maritime contacts that it had before the advent of the railway. Kirkcudbright, Wigtown and Stranraer were all medieval burghs engaged in the coastal trade although only Stranraer retains its importance as a port and regional centre.

The Isle of Man is a self-governing community which, although not represented in the British Parliament, has been influenced throughout its recent history by Britain. Geologically, it is essentially a block fault standing up as an enclave from the submerged Irish Sea and is the link between Cumbria, North Wales, Wicklow and Northern Ireland.

The island can be very broadly divided into the uplands, plateaux and benches of the south and the very flat northern drift. There are two main upland areas which can be seen on the image as brown patches representing the moorland vegetation. The southern area, centred on South Barrule, rises to a height of 485 m whilst centred on the northern area is Snaefell (620 m). The moorlands are mainly grass-covered except at higher altitude where bare rock appears. Flanking the upland areas are parts of an old dissected platform covered by a variety of soils which have supported farming activities since ancient times.

The other main area is that of the northern drift lowlands where the soil is the deepest and best on the island and the climatic conditions most favourable. The chief crops are oats, roots, potatoes and some barley. To the extreme north can be seen the light area representing the sand flats and shingle of the Point of Ayre.

Douglas is by far the largest settlement on the island. Its development was principally due to its steamer link with mainland Britain and its position on the more sheltered eastern side of the island.

= 5 km.

17. Cumbria

For reference see O.S. Routemaster Series, Sheets 3 and 5

Am Ambleside
An Annan
As Aspatria
Ba Barnoldswick
BL Broomlee Lough
Br Brampton
BR Burnhope Reservoir
Bu Buttermere
Ca Carnforth
Ck Cockermouth
Cl Clitheroe
CM Cleator Moor
Co Colne
CW Crummock Water
Da Dalton-in-Furness
Eg Egremont
EW Esthwaite Water
G Grasmere
GL Greenlee Lough
Gr Grange-over-Sands
Gt Gretna
Ha Haweswater
Ke Keswick
KR Killington Reservoir
LK Loch Kinder
LW Loweswater
Ma Maryport
Mi Millom
MT Malham Tarn
Sl Silloth
SR Stocks Reservoir
U Ullswater
Ul Ulveston
W Windermere
Wi Wigton

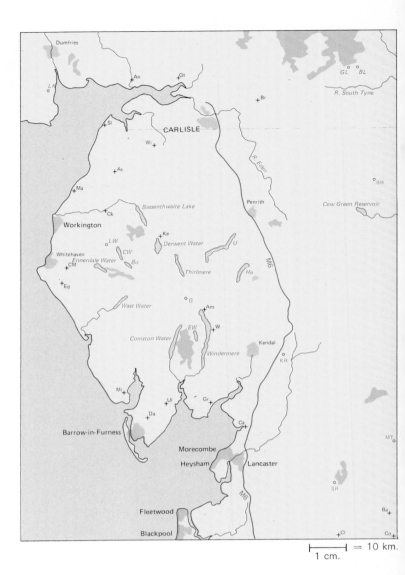

$\vdash\!\!\!\!-\!\!\!\!-\!\!\!\!-\!\!\!\!\dashv$ = 10 km.
1 cm.

Cumbria stretches southwards from the Solway Firth to Morecambe and eastwards from the Irish Sea to the Pennines. Its general characteristic is that of a central upland dome surrounded by lowlands.

This image can be divided into a number of regions based upon the physical structure of the area. The most obvious feature is the light coloured Cumbrian dome (the Lake District) with its elongated radiating lakes. The entire area was covered with new red sandstone and in Tertiary times arched into a dome shape. The drainage pattern was initiated before the sandstone was removed so that the upper courses of the rivers are in the old rocks whilst the lower courses are in the newer ones. The Lake District has a modified form of radial drainage, related to the east/west watershed. This appearance of radial drainage is accentuated by the pattern of the lakes related to glacial overflow from the former ice field centred on the Lake District. The lakes now present in this area may either have been due to glacial over-deepening (Wast Water and Haweswater in particular) or morainic barriers.

Surrounding the central dome are the bright green areas of the southern fringe, west Cumberland and the Solway Plain. The southern fringe extends eastward from Barrow-in-Furness around Morecambe Bay to meet the main Lancashire Plain. Extensive areas of sand may be seen in Morecambe Bay. Along the western coast, centred around Maryport and Aspatria, is the Cumberland coalfield. Iron ore, obtained from the carboniferous lime-stone series, led to the iron and steel industry, now concentrated at Workington, Millom and Barrow-in-Furness. To the north lies the Solway Plain which has extensive areas of arable land, where oats, swedes and potatoes are grown.

Running in a south-east direction from the Solway Firth is the Eden Valley, forming a transitional zone between the Lake District and the Pennines. This is a natural route between Scotland and England, separated from the Lancashire coastal plain by Shap Fells. Before electrifica-tion of the main West Coast railway this formed a severe obstacle to rail travel. The M6 motorway also follows this route and can be seen on the image winding through the Lune Gorge prior to ascending Shap.

The true division, however, between Cumbria and the Pennines consists of a series of geological faults. The Pennine fault is the most marked, producing a steep scarp overlooking the Eden Valley. This is marked by an almost straight line where the green of the Eden Valley meets the acid, peaty vegetation covering the grits which overlay the uplifted limestone of the Alston block. This is separated from the carboniferous limestone of the Askrigg block by the Stanmore trough, seen where the green of the Eden Valley nearly meets the eastern edge of the image. The Askrigg is lighter than the Alston block owing to differ-ences in vegetation due to better drainage and a less extensive covering of grit. In areas where the millstone grit caps the limestone, e.g., Wernside, the peaty surface cover shows up as a darker brown.

18. North Yorkshire and Teesside

For reference see O.S. Routemaster Series, Sheets 4, 5 and 6

Bi   Billington
Br   Brotton
Ca   Carrville
GD   Great Driffield
He   Hetton-le-Hole
Ho   Hornsea
HS   Houghton-le-Spring
Ja   Jarrow
Kn   Knaresborough
Ma   Marske-by-the-Sea
Mu   Murton
MW   Market Weighton
No   Norton
Pi   Pickering
Po   Pocklington
S   Seaton Delavel
Se   Sedgefield
St   Stokesley
Th   Thirske
Wa   Washington
Wh   Whitburn

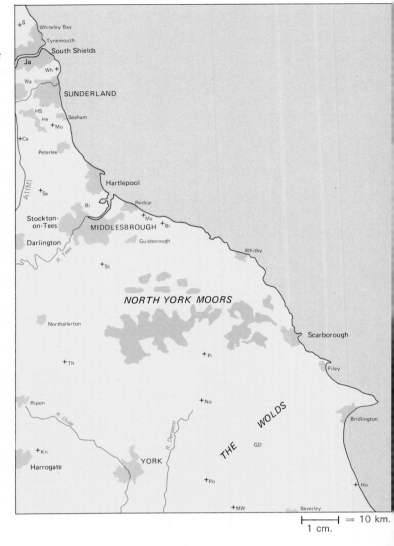

The most striking feature that can be seen on the image is the dark brown area representing the upland regions of north-east Yorkshire, the North York Moors. These hills consist mainly of the sandstones, grits and shales of the middle Jurassic period. Peat and heather moorlands are found on the impervious sandstones whilst bracken and woodland cover the steep slopes of the shales. Large parts of the area are being put down to conifer woodland, particularly to the west of the region on the Corallian sands. Traditionally this was an area associated with sheep farming but during recent years this has declined owing to an increase in dairy farming.

The oldest and most westerly town of the Teesside conurbation is Stockton-on-Tees. The industrial expansion of this area, one of the most rapid in Britain, was centred on the iron and steel industry. Its principal advantage lay in the close proximity of the tidal waters of the River Tees to the ores being worked in the Cleveland Hills. Marine engineering formed a link between this industry and the shipbuilding of the Tyne and Wear, although this is now in decline. The chemical industries of Teesside, evolved from salt production derived from brine wells around the estuary, is now one of the most important industries.

In the south-eastern portion of the image may be seen a bright green area representing the chalk of the Yorkshire Wolds. This can be seen curving inland from Flamborough Head (to the north of Bridlington), south towards Market Weighton. The Wolds have scarpfaces to the north and west, overlooking the low lying clay vales of Pickering and York whilst the dip slope falls away towards the east coast. This forms the region of Holderness, where glacial and post-glacial deposits cover the chalk. Here the image can be picked out in a darker green, in the extreme south-east corner. The High Wolds have thin, dry soils that have been traditionally given over to rough sheep grazing. Rotational farming methods have improved the soil characteristics and now a variety of crops are grown to supplement sheep farming, which still retains its importance. The main towns of the area are Bridlington, Great Driffield, Beverley and Pocklington.

Lying between the higher lands of the Wolds and the North York Moors is the Vale of Pickering. This can be seen on the image as a lighter strip running between the dark areas of the moorlands and the brighter green of the chalklands. The feature is almost completely enclosed, being cut off from the Vale of York by the Howardian Hills through which the River Derwent has cut a 60 m gorge. During the ice age, ice dams to the east caused the formation of Lake Pickering, this accounting for the very flat floor of the Vale.

Extending northwards from the southern edge of the image are the broad lowlands of the Vale of York. These taper beyond the North York Moors to join the Tees Valley, thereby forming a natural link between the industrial areas of Teesside and South Yorkshire.

= 5 km.

19. Northumbria and Teesside

For reference see O.S. Routemaster Series, Sheets 4 and 5

Am   Amble-by-the-Sea
B   Billingham
BC   Barnard Castle
Bi   Birtley
Bl   Blaydon
BL   Broomlee Lough
Br   Brotton
BR   Burnhope Reservoir
Ca   Carrville
Ch   Chester-le-Street
Co   Corbridge
Cr   Crook
CR   Catcleugh Reservoir
Fe   Ferryhill
GL   Greenlee Lough
He   Hetton-le-hole
HS   Houghton-le-Spring
Hx   Hexham
Ja   Jarrow
KS   Kirkby Stephen
Ma   Marske-by-the-Sea
Mu   Murton
Nb   Newbiggin-by-the-Sea
Pr   Prudhoe
S   Sedgefield
Se   Seaton Delavel
Sh   Stanhope
St   Stokesley
Th   Thirsk
W   Whickham
Wa   Wallsend
Wh   Whitburn
Wi   Willingham

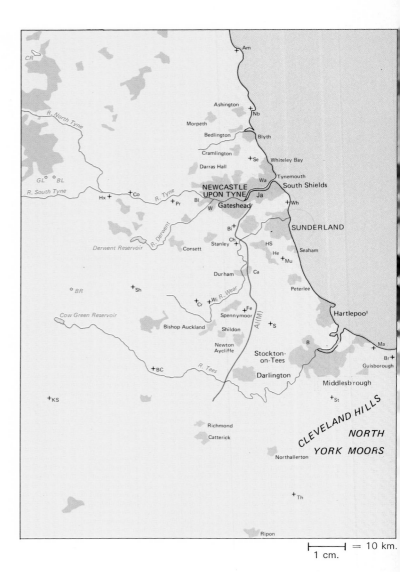

$\vdash\!\!\!\!\!\dashv$ = 10 km.
1 cm.

This image stretches southwards from the Cheviot Hills to the Vale of York and the North York Moors and eastwards from the Pennines to the North Sea. The Pennine chain, of which the Cheviots form the northern-most part, can be seen as the dark brown areas. Along the coast are the light brown areas representing the large industrialized conurbations of the north-east, whilst to the south of the image is the lighter area of the Vale of York, which narrows through the Northallerton gap to join the Tees Valley.

The Cheviot and Northumberland hill zone is separated from the Alston and Askrigg blocks of the Pennines by a system of faults. This may be seen clearly on the image as the narrow green strip, running in an east/west direction through Hexham, separating the brown upland areas. The rocks of the Northumberland hill zone are mainly carboniferous sediments, with extensive igneous intrusions, such as the Cheviots and the Great Whin Sill. The area is covered by moorland, but in the lower eastern slopes sheep and cattle are reared. In the north-west of the image may be seen the Kielder Forest.

The lower lying region, sandwiched between the Pennines and the coast and extending from the top of the image south towards Northallerton, has been termed the North-East. The areas of industrialization coincide closely with the limits of the Northumberland and Durham coalfield. There are three major areas of settlement, the Tyne, Wear and Teesside conurbations. The largest of these is Tyneside, stretching from Newcastle and Gateshead along the Tyne valley and north and south along the coast. Although strictly speaking the conurbation only extends northwards to Whitley Bay, the almost continuous brown of the built-up region covers an area as far north as Ashington. Newcastle-upon-Tyne, the centre of the area, was an old fortified bridging point that owes its current importance to the growth of Tyneside as a commercial waterway. The main products of the Tyne were shipbuilding and coal shipment from the adjacent coalfields. The Tyneside shipyards specialized in naval shipbuilding prior to World War I, a factor that was very influential in their decline in the years immediately afterwards.

The second major conurbation at Wearside is smaller than Tyneside although equally well defined. It is a centre for the colliery settlements of Durham. Because of the rocky nature of the Wear channel, it could not be dredged as thoroughly as the Tyne and consequently much of the coal traffic was diverted from the interior towards the Tyne rather than the Wear. Shipbuilding and marine engineering are the main industries.

Teesside, whose development was linked with the iron and steel industry, is the major metallurgical centre of the North-East now that the Consett Steel Works have been closed. The area also has an important chemical industry at Billingham producing ammonia, fertilizers, acids and crude petroleum, whilst at Teesport an oil refinery and petrochemical works was constructed during the 1970s.

20. The North Channel

For reference see O.S. Routemaster Series, Sheet 3

Bc Ballycastle
Cu Cuchendale
LG Loch Glashan
PB Port Bannatyne
Ro Rothesay

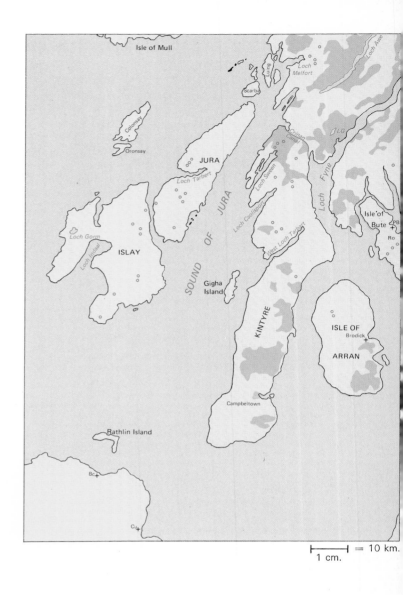

├─────┤ = 10 km.
1 cm.

This scene embraces the islands of Islay and Jura of the Inner Hebrides, Kintyre and the approaches of the Clyde, together with Ulster, separated from Scotland by the North Channel. The entire area has been extensively glaciated and much of the sea area is formed of sea-floor basins. Major basins lie between Jura and Mull (to the top of the image) and Arran and the mainland. Much of the glacial erosion was influenced by the geology of the area. For example, the channel between Colonsay and Jura is probably related to the presence of relatively weak Mesozoic strata on the sea-bed.

The two most easterly islands, Arran and Bute, are served by regular steamers from the Clyde and Ayrshire coast, and are centres for tourism. Evidence of the more hostile terrain of Arran may be deduced from the image, where the brown areas represent extensive peat moorlands. The island is very rugged and mountainous, rising to a height of approximately 900 m. Most of the farming settlements are found on the alluvial flats and raised coastal beaches that may be seen as the lighter areas on the image. Arable farming is practised in these areas, whilst the higher regions are given over to rough grazing for sheep. On the lower and flatter Bute, dairy farming and market gardening are important, the produce being shipped through Rothesay to the nearby industrial Scottish Midlands.

Running southwards from the top of the image is the long peninsula of Kintyre. This is considered to be part of

the Grampian region and although generally lower, the northern part shows evidence of the dissected appearance associated with this area. This takes the form of parallel ridges and hollows produced by ice action, that exploited the contrast in resistance of the northeast, southwest lying rocks. Examples may be seen about Loch Sween on the west coast of Kintyre. During the last ice age many areas were fragmented by glacial breaches. Surrounding the peninsula are numerous raised beaches which are extensively farmed for winter feed for sheep and dairying. In the low-lying trough connecting Campbeltown to the western coast, the Argyll coalfield outcrops, at one time providing a thriving export trade to Northern Ireland. This main settlement of the peninsula is a herring fishing centre and the site of several distilleries. Towards the north of Kintyre, connecting Loch Fyne to the Sound of Jura, can be seen the Crinan Canal. This was built during the early 19th Century as a means of improving communications between north and west Scotland with the Clyde and west central Scotland.

Islay, Jura and Colonsay are part of a group of islands known as the Inner Hebrides. Islay, Jura and the smaller islands of Scarba and Luing probably formed a continuous peninsula of the mainland during pre-glacial times. Their separation was due to glacial breaches, whilst further fragmentation nearly took place on Jura which was almost cut in two at Loch Tarbert.

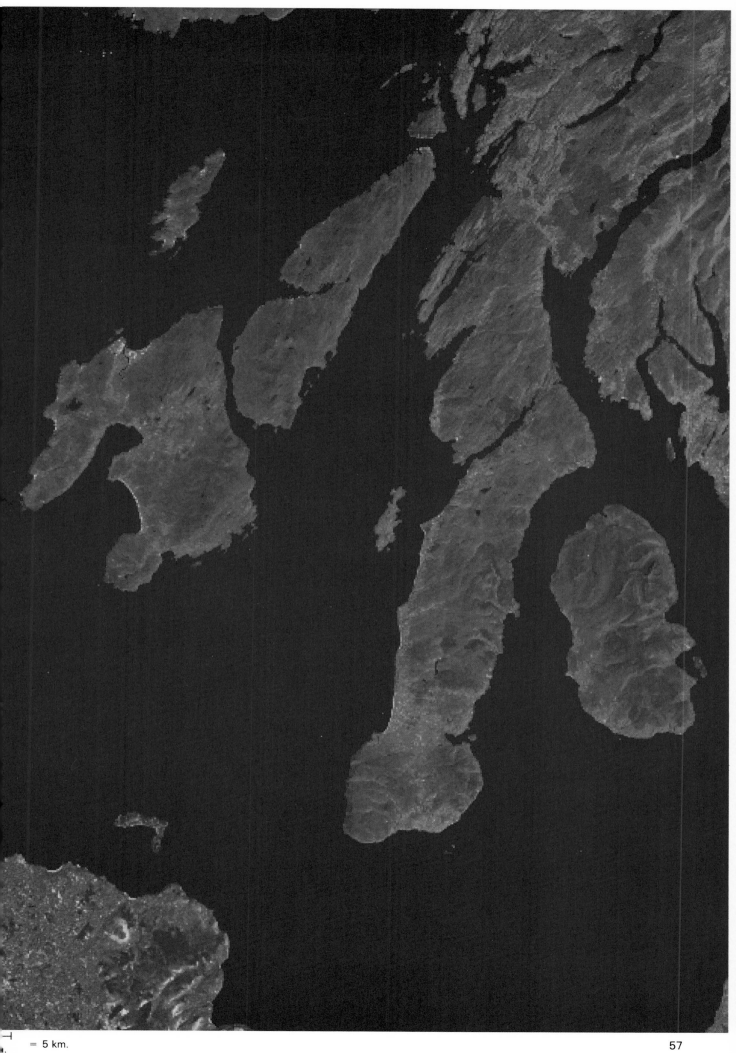

= 5 km.

21. The Ayrshire Basin and Southern Uplands

For reference see O.S. Routemaster Series, Sheet 3

Al   Alexandria
Ar   Armadale
Au   Auchterardy
Ba   Bathgate
BA   Bridge of Allan
Bo   Bonhill
C   Clydebank
Ca   Callander
Cl   Carluke
Dr   Darvel
Ho   Holmhead
Hu   Hunterston
In   Innellan
Jo   Johnstone
Kb   Kilbirnie
Ki   Kilsyth
LA   Loch Ard
LAR   Loch Arklet Reservoir
LBR   Loch Braden Reservoir
Lh   Larkhill
LL   Loch Lubnaig
LM   Lake of Monteith
Ln   Lanark
LS   Loch Sloy
LV   Loch Venachar
Nm   Newmilns
Pa   Paisley
Ru   Rutherglen
St   Stewarton
Wh   Whitburn

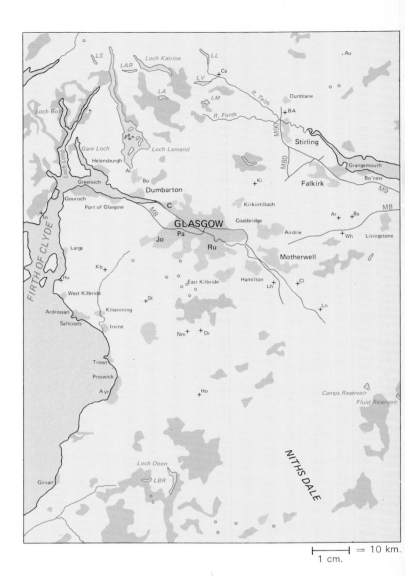

├──────┤ = 10 km.
1 cm.

This image can be sub-divided into a number of easily identifiable regions. To the north may be seen the Highland zone with its pronounced valleys and lochs. Further south may be seen the volcanic hills, whilst running across the centre of the image are the industrialized Central Lowlands with their extensive urban areas. Along the western coast is the Ayrshire Basin encircled and separated from the Central Lowlands by higher ground. The remaining area consists of the Southern Uplands bisected by Nithsdale.

The Ayrshire Basin is an amphitheatre of lowlands enclosed by a wide area of hills and moorland drained by rivers running towards the Firth of Clyde. (They are not shown on the map but may be located on an Ordnance Survey map.) Where the higher ground meets the sea, near Largs and Girvan, a narrow coastal strip is terminated abruptly by steep hills. The lowlands give rise to higher ground which forms a watershed between the Clyde and the Ayrshire Basin. It is possible to see gaps in these hills that form important communication links. South of Darvel, situated towards the centre of the image, the hills become higher and merge with the Southern Uplands. Ayrshire is renowned for its high quality dairy products and early potatoes. The damp, mild climate favours the growth of oats and fodder crops although about 60% of all good quality land is grassland. Much of the milk produced goes to the industrial towns of Ayrshire and the central lowlands, although an increasing amount is processed into butter, cheese and dried milk.

One of the major industries is coalmining, now centred around Holmhead, production being concentrated in new or reconstructed collieries. The coal is used for both industrial and domestic purposes and significant quantities are shipped to Ireland and the Hebrides through Ayr, Troon and Ardrossan. Prominent features that can be identified include the international airport at Prestwick and the nuclear power stations and deep water ore terminal at Hunterston. Many of the coastal towns in this area support a large tourist industry owing to their close proximity to the large population of Clydeside.

To the south-east of the Ayrshire Basin are the Southern Uplands. To the east of Nithsdale is a region known as the Central High Plateau, whilst to the west are the Galloway Uplands. The bright green of Nithsdale is very prominent and unlike the other valleys of this region is very broad and low. The Galloway Uplands are very dissected and the uniformity of the grits and sandstones is broken by granitic intrusions. A large area of granite occurs to the south and west of Loch Doon and may be seen on the image as an area of darker brown, slightly lighter than the areas of woodland. The climatic conditions and soil cover are suited to spruce and since 1930 large Forestry Commission plantations have been established.

22. The Central Lowlands of Scotland

For reference see O.S. Routemaster Series, Sheet 3

A   Armadale
Ar   Auchterarder
Au   Auchtermuchty
Ba   Bathgate
BL   Bonnyrigg and Lasswade
Ca   Callander
Cl   Clydebank
Co   Cowdenbeath
CVR Carron Valley Reservoir
Dk   Dalkeith
Dr   Darvel
In   Inverleithen
Jo   Johnstone
Ki   Kilsyth
LA   Loch Ard
LBR Loch Bradan Reservoir
Lh   Larkhill
LM   Lake of Monteith
Ln   Lanark
Lo   Lochgelly
Lu   Loch Lubnaig
LV   Loch Venachar
Nb   Newburgh
Nm   Newmilns
Pa   Paisley
Pe   Peebles
Pp   Preston Pans
Qu   Queensferry
Ru   Rutherglen
St   Stewarton
Wt   Whitburn

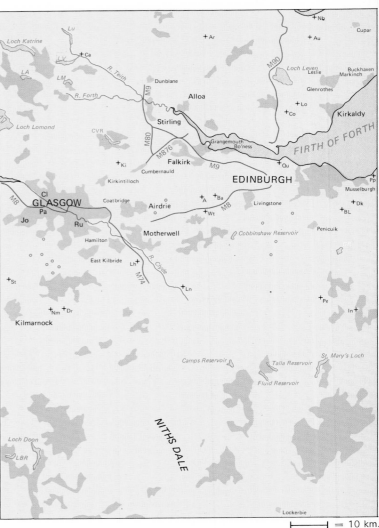

1 cm. |⊢————⊣| = 10 km.

The Central Lowlands are accepted as a geographical entity and are formed by a downfaulted rift block between two major fault lines, the Highland boundary fault to the north and the Southern Uplands fault in the south. Unlike the higher masses to the north and south, the Central Lowlands consist largely of relatively recent sedimentary deposits interrupted by igneous intrusions and masked in many places by generous deposits of marine, fluvial and glacial deposits. With its lower surface, more varied soil and favourable climate, this is the economic heartland of Scotland. Approximately 75% of Scotland's population live in this region, 15% of the country's area. This area is clearly defined on the image and extensive built-up areas around the Upper Clyde and Falkirk are easily identifiable. The reasons for the great industrial concentrations are numerous; lower altitudes and productive soils encouraged farming; large coal resources and iron ore deposits provided the basis for heavy industries whilst the penetration of the area by long, deep water estuaries gave access to shipping.

Running in a roughly north-west direction across the centre of the image is the Clyde Valley. Although thought of as an industrial area, farming is the main occupation of most of this region. The climatic conditions and the requirements of the industrial conurbations have led primarily to the production of milk, beef and root crops. The industrial impetus was given by the readily accessible coal in the Lanarkshire coalfield coupled with the interbed-ded ironstone. This, combined with the double outlook to the Atlantic and the North Sea, has contributed principally towards the industrialization of the Central Lowlands. Coal is mined throughout the region and the coalfields extend eastwards across the Firth of Forth to beyond Kirkcaldy, and also outcrop in the Lothian region to the east of Edinburgh. The dominant industries of this region are iron and steel, shipbuilding and engineering.

On the southern shore of the Firth of Forth is Edinburgh, which originated as a small burgh built about the castle rock. Its strategic position, commanding the main coastal route from England, and its cultural links with the continent, led to it becoming the centre of government, administration and law. It stands in an area of rich agricultural land and has important engineering industries.

South of the Lowlands surrounding Edinburgh is the high Central Plateau of the Southern Uplands. Running in a north-east/south-west direction it is possible to identify a division that corresponds to the location of the Southern Upland fault. Communications have always been difficult across this area and the valleys are the easiest routeways. The principal route between Carlisle and Glasgow for both road and rail is via Beattock Pass. The A74 carries approximately 75% of the road traffic between England and Scotland. This route can be identified running in a north-north-easterly direction from Lockerbie.

= 5 km.

23. The Cheviots and the Merse of Tweed

For reference see O.S. Routemaster Series, Sheet 4

Am Amble-by-the-Sea
An Anstruther
Co Coldstream
CR Catcleugh Reservoir
Cr Crail
Du Duns
Ey Eyemouth
Ke Kelso
Me Melrose
Nb Newbiggin-by-the-Sea
Pi Pittenween
Se Selkirk

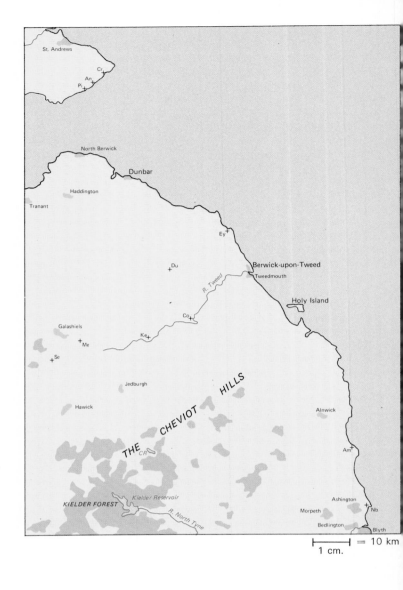

= 10 km

1 cm.

The Cheviot Hills, seen in the south of the image, represent the highest portion of the Pennine Chain, rising to 815 m. They consist largely of Devonian lavas later intruded by Tertiary granite. Erosion has removed the early basalt and left the granite as flat topped hills with steep but rounded sides. This can be seen on the image as the light brown area to the south-west of Holy Island. Surrounding the igneous intrusions are large areas of sandstone, with thinly interspersed limestone beds. The tops of the Cheviot Hills are covered with peat bogs which thin out along the lower slopes. The valley floors have deposits of boulder clay, sands and gravels of glacial origin and have good short grass cover, for stock raising of sheep and cattle. The main route across these hills is the A68 which converges with the A696 north of Otterburn, to ascend Redesdale and the pass of Carter Bar to Hawick and Jedburgh. It is possible to follow this on the image. In the south-west are the large, dark brown areas of Kielder Forest and Redesdale. Forestry is now an important occupation in this area, where extensive plantations of Sitka and Norway spruce have been established on former open sheep runs. In the midst of Kielder Forest is Kielder Reservoir, one of the largest man-made reservoirs in Britain.

North of the Cheviots is the Tweed Basin. Centred on Coldstream this clearly defined basin is enclosed by a highland crescent on three sides but opens eastwards to the sea and the main east-coast routes. During the last ice age, boulder clay, sands and gravels were deposited on the floor of the basin. Rainfall is evenly distributed throughout the year and the summer temperatures are higher than elsewhere in Scotland. The area is intensively farmed, and farms are highly mechanized. The principal crops are grass, roots, oats, wheat and barley, while the Tweed Basin generally carries one of the highest concentrations of sheep in Britain. Store cattle are also important. The tradition of sheep farming was established in this area by the Cistercian monks in the abbeys at Kelso, Melrose and Jedburgh. This in turn led to the establishment of a textile industry which expanded during the 18th century. A rapid expansion of the industry was seriously handicapped by the lack of readily available coal which proved to be a serious disadvantage in the face of competition from the coal-based Yorkshire textile industry. The principal towns of the area are Hawick and Galashiels, both famed for textile production, and Berwick-upon-Tweed.

To the north of the Tweed Basin are the Lammermuir Hills which are marked by a well-defined north-facing escarpment. This is the most conspicuous portion of the Southern Upland fault, forming the southern boundary of the Central Lowlands and can be seen on the image.

24. The Sea of the Hebrides

For reference see O.S. Routemaster Series, Sheets 1 and 3

LC   Loch Coruisk
LD   Loch Druidibeg
LP   Loch a' Phuill

Within this image two separate groups of islands can be defined: the Inner Hebrides on the eastern part of the image separated from the southern parts of the Outer Hebrides by the Sea of the Hebrides. Throughout the area rainfall is lower than on the mainland, owing to the low altitude of the islands, and the mild winters. Most of the area is given over to grass and rough grazing and in the Hebrides there are approximately half a million sheep.

The geology of the Inner Hebrides is varied, but the most important feature is the variety of complex igneous intrusions and outcrops arising from the Tertiary period. These have since been denuded to form low broad hills and plateaux that rarely exceed 600 m. The area has been extensively affected by glacial action and ice from the mainland has cut deep rock basins, e.g., Loch Coruisk on the Isle of Skye. Soils vary from island to island, but usually consist of sand, gravels and pebbles on the raised beaches with peat in the higher areas. Tiree and Coll, however, have beaches covered with fertile calcareous loess and are almost peat free, hence the light green colour on the image. Generally the chief industry for the islands of this group is agriculture within the crofting system, with a predominance towards sheep farming. Tiree was originally developed as part of the Duke of Argyll's estates. This led to the area being divided into consolidated holdings, crofts, of 1·5—4 ha with the pasture land attached to the crofts becoming common grazing for the crofting townships. The crofting system was designed as a part-time farming unit with a seasonal occupation, such as kelp manufacturing or fishing. This form of reorganization of farms into crofting townships spread rapidly throughout the north and west during the first two decades of the 19th century. Tiree, with its lime-rich, sandy deposits is the most progressively farmed island. Government assistance has been given towards the development of the bulb industry and the glasshouse cultivation of tomatoes.

The Sea of the Hebrides, separating the two groups of islands, is a rock basin formed by ice action. The most remarkable feature of this channel is a major trench approximately 10 km wide and 100 km long, that extends south-westwards from near the island of Rhum. The rockhead, as distinct from the silted bottom, is 300 m and in some areas 380 m below sea level.

In the north-west of the image are the most southerly islands of the Outer Hebrides, Barra and South Uist. The relief is generally subdued, largely due to extensive ice action, but is diversified by isolated hills that often rise abruptly from it, such as Hecla and Beinn Mhor in South Uist (approximately 650 m). Much of the area consists of huge slabs of bare rock alternating with boggy, peat hollows and numerous lochans in a virtually treeless landscape. The major occupation in this area, apart from crofting, is fishing with a concentration of activity at Barra.

= 5 km.

25.   The Great Glen Fault

For reference see O.S. Routemaster Series, Sheets 1, 2, 3 and 4

LA   Loch Affric
LAr   Loch Arienas
LB   Loch Beinn a' Mheadhoin
LBe   Loch Beoraid
LDa   Loch an Daimh
LE   Loch Eilt
LEM   Loch Eilde Mór
LL   Loch Leathan
LLo   Loch Long
LLu   Loch Lundie
LNa   Loch Nant
LO   Loch Oich
To   Tobermory

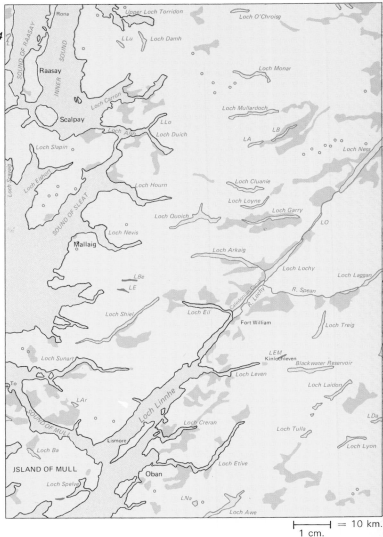

⊢————⊣ = 10 km.
1 cm.

The most pronounced feature on this image is the Great Glen fault separating the north-west Highlands from the Grampian Region. The Great Glen (or Glen More) fault is one of the world's great topological schisms and is now occupied by a series of long narrow deep lochs, stretching over 80 km from Inverness to Fort William and beyond into Loch Linnhe. It is a tear fault with lateral shifting. Earthquake activity of varying intensity has been recorded in this area (e.g., in 1816 and 1901). Its formation caused alterations to the drainage pattern whilst glaciation produced deep rock basins such as Loch Ness.

The entire area of the image comprises the highest and least populated part of Britain. Composed largely of crystalline and metamorphosed rocks rising in parts to over 1300 m, the Highland massif is one of considerable geological complexity. The grain of the country generally follows the characteristic north-east/south-west Caledonian trend. The north-west Highlands are great rugged mountain blocks separated by deep, narrow glens, clearly seen on the image. Most of the fiord floors of western Scotland descend seawards to a depth of 100–200 m before the characteristic rise at their entrance. This descent is often irregular, resulting in alternating rock bars and basins. Loch Etive has two such basins. The rock bars are usually submerged but sometimes form marked constrictions, e.g., Upper Loch Torridon, whilst that at the mouth of Loch Etive is so shallow that it produces a reversing tidal waterfall (the Falls of Lora). Much of the north-west Highlands are barren, the valleys occupied by long lochs and stretches of bogland, whilst between them are wet, peaty swamps and moors at heights of 250–450 m. The annual rainfall is over 2500 mm; snow falls on an average of 40 days in a year and the frost-free period is less than 90 days. As the summers are cool and rainy the land has little opportunity to dry out. Coupled with a soil that contains few plant nutrients, it is only on the flat, relatively sheltered valley floors where any cultivation is possible. The more favoured areas, with a cover of rough grass and heather, permit sheep farming, but the density is very low, one sheep to about 10 ha. High rainfall combined with steep slopes and the pattern of the valleys have helped make this an important area in the development of hydro-electric power. Most of the larger valleys either have schemes planned, under construction or in operation.

The Grampians show greater variety of scenery due partly to a west/east change from metamorphic to igneous rocks but also because of the degree of weathering. The granite intrusions produce high plateau areas such as Rannoch Moor (to the south-east of Kinlochleven) whilst the Caledonian trend is shown in the grain of the country and in the direction of the faults. Loch Etive follows one such line of weakness, excavated later by rivers and ice. This area, too, is important for the generation of hydro-electric power, pioneered by the British Aluminium Company in 1896. Aluminium is still produced in large quantities at Fort William and Kinlochleven.

= 5 km.

67

26. The Grampian Mountains

For reference see O.S. Routemaster Series, Sheets 2 and 4

Ab   Aberfeldy
Al   Alyth
BR   Blackwater Reservoir
Br   Braemar
Ch   Charlestown of Aberlour
Co   Coupar Angus
Gr   Grantown-on-Spey
LA   Loch Ashie
LAg   Loch Aggon
LDa   Loch an Daimh
LEr   Loch Errochty
LFr   Loch Freuchie
LLi   Loch of Lintrathen
LR   Loch Ruthvern
LV   Loch Voil
MO   Muir of Ord
Mo   Monifieth
N   Newport-on-Tay
Ne   Newtonmore
NS   New Scone
Ra   Rattray
Ro   Rothes

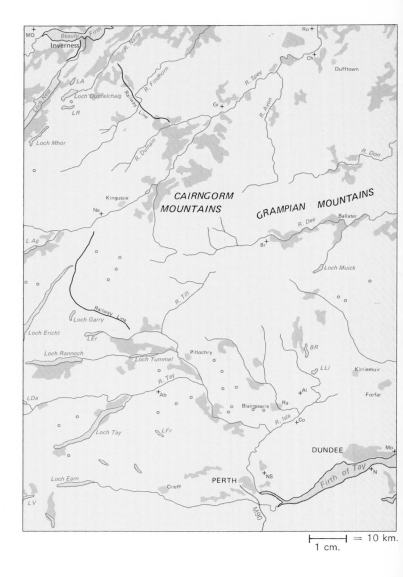

$\longmapsto$ = 10 km.
1 cm.

This image not only portrays the great mass of the Grampian Mountains, sandwiched between the Great Glen Fault in the north-west and the Highland Boundary Fault to the south-east, but also emphasizes the characteristics of the region. These characteristics alter east and west of a line joining Inverness to the Tay Valley. To the west, the Grampians are dissected with pronounced ridges and peaks and a large number of elongated lochs, whilst to the east, the mountains are broader with flat, moorland tops cut by deep glens with few large areas of water. Glaciation has been mainly responsible for these differences, since it occurred more than once in Scotland with varying degrees of intensity. The first glacial period (the Great Highland Glaciation) formed an ice cap apparently covering the whole area, giving the eastern part of the Highlands their rounded appearance. The second glacial period (The Scottish Glaciation) was less extensive, being confined to a few glaciers which produced the characteristic U-shaped valleys and other associated features. In the west, where the precipitation and quantities of snow are greater, the glaciers cut more deeply, particularly along fault lines and shatter belts. Here the lochs are rock basins rather than dammed valleys, and erosion is very marked where they narrow. Loch Ericht, for example, is 23 km long and never more than 1 km wide, with a deepest point of 156 m.

In the north-west of the image, between Loch Ness and the Cairngorms, are the Monadhliath Mountains. The stepped plateau features generally rise to about 300 m with several peaks of over 750 m (e.g., Carn Ban, 910 m).

Much of the area has been smoothed by erosion and extensively covered by peat and glacial debris. There is no large-scale settlement and much of the land is occupied by grouse moors. To the east of this area, using the Spey Valley, may be seen the railway line from Inverness to the south.

The snow-covered Cairngorm Mountains are near the middle of the image. These are higher and show evidence of more intense glaciation. The main plateau has a height of approximately 900 m but many summits rise from it (e.g., Cairn Toul, 1220 m). Subject to prolonged snow cover and with smooth and lengthy slopes, this area has recently been developed into an organized skiing centre.

Throughout the mountain area of the image, several long wide green valleys can be seen. These lowlands are termed 'straths' and radiate from the highlands towards the fringing lowlands. The most important of these is Strathspey, seen in the north-east of the image running north-eastwards between the Cairngorms and the Monadhliath Mountains. Tourism and forestry are important with crofting and sheep farming in the upper reaches and beef cattle in the more sheltered regions.

In the south-east of the image can be seen the fertile, sheltered region of Strathmore, separated from the Grampians by the Highland Boundary Fault. Oats, potatoes and wheat are cultivated and the area is important for the fattening of Angus cattle. East of Perth, food crops become more important than animal rearing and about 80% of the Scottish raspberry crop is grown here.

27. The Buchan Peninsula

For reference see O.S. Routemaster Series, Sheets 2 and 4

Ba   Ballater
Bc   Banchory
BR   Blackwater Reservoir
CA   Charlestown of Aberlour
Co   Coupar Angus
Cu   Cullen
In   Inverbervie
La   Laurencekirk
LLi   Loch of Lintrathen
LS   Loch of Skene
Ne   Newport-on-Tay
Pe   Peterculter
Po   Portnockie
Ro   Rothes
Wh   Whitehills

├────────┤ = 10 km.
1 cm.

This image depicts the eastern slopes of the Grampians as they slope downwards to meet the North Sea. The eastern tilt of this surface is shown by the courses of the rivers Dee and Don. The core of the area is the Buchan Plateau, the brown/green, north-eastern portion of the image between the Deveron and the Don. It was probably spared the action of glacial ice by the Scandinavian ice off the coast forcing the Scottish ice from the north-western Highlands to divide. This left the Buchan area as one of deposition rather than one of intense erosion. The best soils in the area are the alluviums, fertile loams and clay loams in the Ugie and Ythan basins. Approximately 80% of this area is either cultivated or under permanent grass. The most important farming industry is cattle-raising within a six-year rotation of oats, turnips and grass. The major towns are Peterhead and Fraserburgh, both associated with the now declining herring industry. To the north of Peterhead may be seen the light area representing the new pipeline terminal built to transport North Sea gas.

West of Buchan, the higher Banffshire Plateau has a sloping drift covered surface that is a transitional farming zone between Buchan and the Moray Firth. The area is well cultivated and barley, seed potatoes and sugar beet are grown. The Plateau slopes gently towards the northern coast and numerous fishing ports are located in the many sheltered bays. The chief inland towns are Tuniff, situated where the main road from Aberdeen to Banff crosses the Deveron; Huntley, a former textile centre; and Keith.

South of Buchan are the lowlands of the Dee and Don valleys. The soils of the Dee valley are less fertile than those of the Don and consequently a greater proportion is wooded. The lower Don flows through good agricultural land where market gardening and the growing of oats, barley, roots and grass are important. Inverurie was once the railway works of the Highland Railway, and is still one of the most important towns in this area. Its other industries include flour milling and paper making.

Aberdeen, at the mouth of the Dee and close to the estuary of the Don, is the only city in north-east Scotland. It originally had links with Flanders and the Baltic ports and became important for the manufacture and export of woollen cloth. Its position opposite Scandinavia also led to a shipbuilding industry, but without the coal and iron these industries did not prosper in the face of competition during the industrial revolution. Aberdeen's current industries include paper making, marine engineering and fishing. The North Sea oil industry has given the city great importance as its headquarters and as a result extensive improvements have been made to the harbour and airport.

In the southern portion of the image can be seen the well defined Highland boundary fault which meets the coast in the region of Stonehaven. The transition between the Highland zone and the lower Strathmore corridor is very abrupt and marked.

1 cm. = 6 km.

28. The Outer Hebrides

For reference see O.S. Routemaster Series, Sheet 1

LB   Loch Brievat
LC   Loch Caravat
LF   Loch Fada
LG   Loch Grunavat
LH   Loch Huna
LL   Loch Langavat
LLe   Loch Leathan
LOb   Loch Obisary
LS   Loch Suainavel
LSc   Lock Scadavay
LT   Loch Trelaval
LU   Loch Urrahag

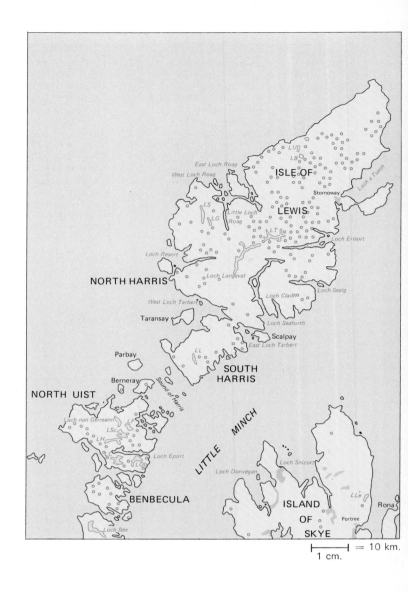

|⊢——————| = 10 km.
1 cm.

This group of islands forms an almost continuous barrier, some 200 km long, between the north-west coast and the Atlantic Ocean. The islands are isolated by the necessity of a sea journey across the Little Minch. The climate is more stormy and cloudy than on the mainland although winters are milder. Lack of sunshine is one of the greatest drawbacks, which when coupled with high rainfall severely limits farming activities. The islands are part of a very old deeply dissected peneplain which reaches a height of approximately 800 m in North Harris. Ice sheets have smoothed the surfaces of the rocks leaving the present landscape as a lake-pitted surface. Submergence of valleys has resulted in the formation of numerous islands, and it is only a narrow neck of land between East Loch Tarbert and West Loch Tarbert that prevented Harris and Lewis from becoming two separate islands. The coastline is heavily indented with narrow, irregular steep-sided fiords and many similar lochs appear within the interior. Glacial drift is widespread throughout the islands and raised beaches of gravel are also common, whilst organic decomposition has produced a thick blanket of peat.

Due to the inhospitable interior, settlements are all close to the sea. Although once a crofting orientated community, beef cattle and sheep rearing are now the most important occupations. One of the most well known industries is the production of Harris tweed, particularly in Lewis. Much of the yarn is produced by mills in Stornoway, from imported raw wool, whilst the weaving still takes place in the home.

The major town is Stornoway, which contains 25% of the population of Lewis and is tending to increase at the expense of the rural areas. This was originally a fishing port but a decline in the fortunes of the fishing industry have resulted in one of the highest unemployment rates in Britain.

In the south-east of the image, separated by the rift valley of the Minch, is the scenic Isle of Skye, one of the most populous offshore islands and also one of the greatest tourist attractions. The island consists largely of a volcanic plateau formed during the Mid-Tertiary Alpine movements. The Cuillin Hills, rising to over 900 m, are composed of the plutonic roots of Tertiary volcanoes.

= 5 km.

29. North-west Scotland

For reference see O.S. Routemaster Series, Sheet 2

LAc   Loch Achall
LBh   Loch a' Bhraoin
LFi   Loch Fiag
LG   Loch Garve
LLe   Loch an Leothaid Bhuain
LM   Loch Morie
LMe   Loch Merkland
LN   Loch Naver
LOi   Loch nah-Oidhche
LOs   Loch Osgaig
LSi   Loch Sionoscaig
LSt   Loch Stack
LV   Loch Vaich
LVe   Loch Veyatie
M   Muir of Ord
Ul   Ullapool

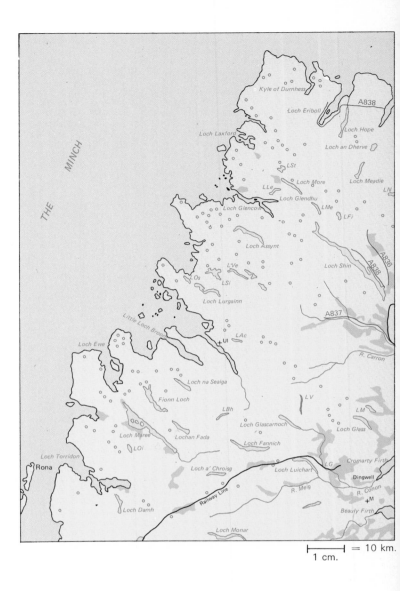

⊢———⊣ = 10 km.
1 cm.

The characteristics visible on this image can be largely attributed to the effects of mountain glaciation, the most striking being the number and form of the numerous fiords that disrupt the entire western coast. Inland, moraine-dammed lochans, ice-smoothed rock, peat filled hollows and thin soil covering the occasional flat piece of land provide poor conditions for economic agriculture. The watershed of this region lies very close to the west coast and the westward moving glaciers exercised maximum effects on their relatively short passages towards the sea.

Most of this region is difficult to cross and contains few major roads and railways. The only railway in the area, to the Kyle of Lochalsh, crosses the southern part of the image, and most roads are restricted to the wider valleys. Along the coast, roads are negligible owing to the narrow, mountain backed coastal plain transversely cut by the numerous sea lochs. The individual communities are very small and tend to be isolated from one another owing to the discontinuous nature of the terrain. They tend to be located at the heads of the fiords, at the mouth of streams or in the wider valleys where flat areas of alluvial soil may be found.

Many of the settlements developed as a result of the Highland clearance in the mid-19th century, when tenants were evicted to allow for the extension of sheep farming and the introduction of deer grazing. Although there was a general exodus towards the industrial south or abroad, many 'clearance villages' grew up and there arose a type of subsistence farming supplemented by fishing and gathering kelp. The flat area of land, on lake deltas or on raised beaches, was farmed for potatoes and oats, whilst stock was grazed on the hill slopes.

This region was recognized as a depressed area during the early 20th century and since then assistance has been given in the provision of new holdings and the enlargement of old ones, and the building of piers and roads. Most crofting settlements still consist of little more than a row of single-storey cottages many of which have been abandoned. Census returns show that more people lived in these regions during the 18th century than do so now. Most young people leave for better prospects in the industrialized areas and many of the older people who remain are unable to cope with the upkeep of the croft and as a result whole settlements are being abandoned.

Ullapool, with road links to Inverness and the Kyle of Lochalsh, is the largest settlement in the area. Recently the tourist trade has supplemented its traditional crofting and fishing activities.

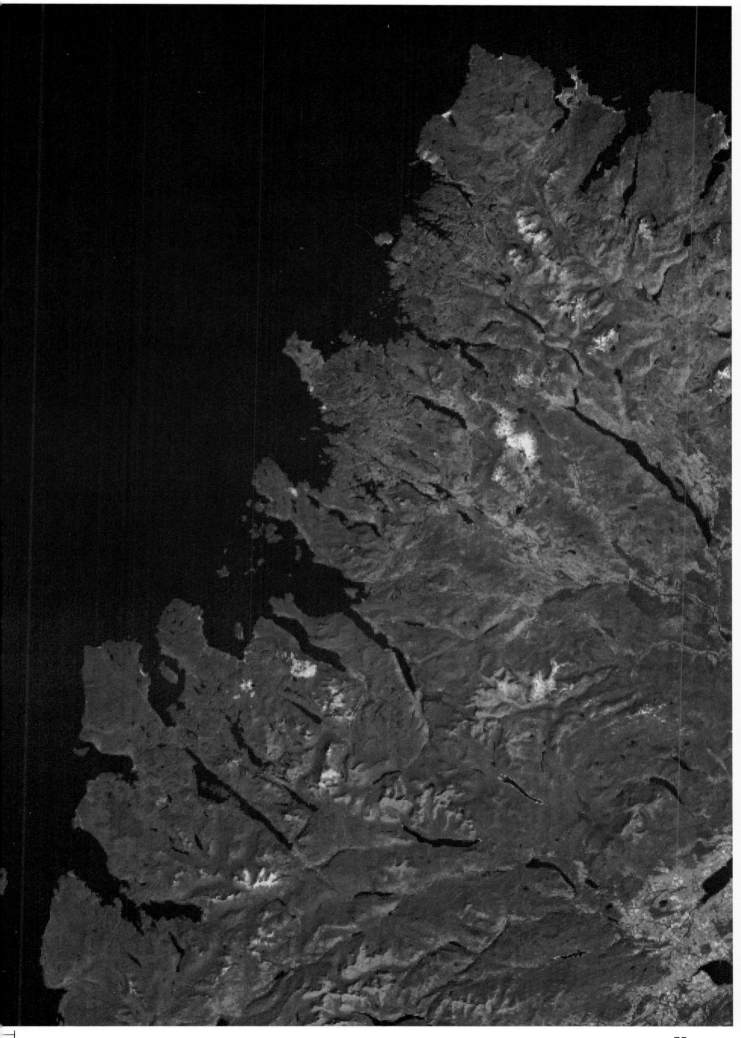

30. Caithness and the Moray Firth

For reference see O.S. Routemaster Series, Sheet 2

Br   Brora
CA   Charlestown of Aberlour
Cr   Cromarty
Cu   Cullen
Go   Golspie
LB   Loch Badanloch
LC   Loch Craggie
LD   Loch Druim a' Chliabhain
LE   Loch Eye
LM   Loch More
LMi   Loch Migdale
LR   Loch Rimsdale
LRu   Loch Ruathair
LS   Loch Shurrery
Ly   Lybster
Po   Portnockie
Ro   Rothes

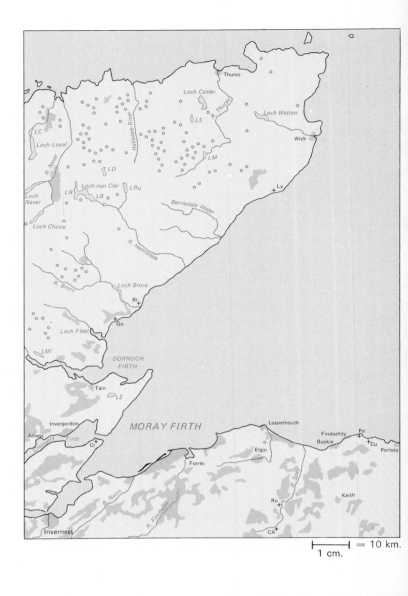

⊢————⊣ = 10 km.
1 cm.

The plateau of Caithness is the eastern part of the north-west Highlands and is an undulating plain of under 150 m in height covered with superficial deposits. The coastline consists of vertical cliffs that have been the subject of severe marine erosion. Connected to the Caithness plateau by a narrow coastal strip, running south-west-wards from Helmsdale, are the lowlands around the Moray Firth. River mouths in this area have been drowned to give the long Dornoch, Cromarty and Beauly Firths. Cromarty Firth is the only deep-water anchorage between the Firth of Tay and Scapa Flow and was once a major Royal Naval Fleet anchorage.

Between the Pentland Firth and Inverness a series of light green lowlands flank the eastern side of the main plateau. In the north-east, the Caithness plain is separated from the Moray Firth by coastal ranges (brown), causing the railway line from Helmsdale to Wick to take a circuitous route through Strath Kildonan. Conditions on the low, north-eastern plateau are favourable for agriculture. The mean annual rainfall is less than 800 mm whilst owing to the maritime conditions, winter temperatures are similar to those of the Isle of Wight. Soils developed from glacial debris and river alluviums are suitable for root crops, barley and oats and cattle and sheep are also reared. The north-east compares favourably with the north-west (Image 29) and there are more large settlements. Wick and Thurso are the most important settlements and both have benefited from the construction of the nearby Dounreay nuclear power station.

South of Helmsdale is the region of the firths, containing the Black Isles and Tarvar peninsulas, where subsidence caused by faulting has created these areas of water. Boulder clay and fluvio-glacial soils cover most of the region and climatic conditions make this the most northerly wheat growing area. Isolation from main centres of population makes beef cattle more important than dairying. This area is becoming important through the North Sea oil programme, and a dry dock has been dug in the sand of Nigg Bay in Cromarty Firth, to accommodate oil tankers up to 500 000 tons.

The eastern lowlands of the Moray Firth consist of an underlying structure of old red sandstone, but the headland to the west of Lossiemouth consists of new red sandstone. Extensive areas of sand dunes can be seen and the westward drift along the coast has caused the formation of sandspits. Considerable problems have been caused by windborne sand which has covered much of the arable coastal land with dunes. Extensive afforestation has to some extent halted the drift. Good soils coupled with a favourable climate have produced an almost continuous arable area. The chief branch of farming is stock rearing, together with rotational crops such as turnips, grasses and oats. Inverness, near to the mouth of Loch Ness, is the largest town of the area. Surrounded by fertile lands, it is the centre of the agricultural area and regarded as a lowland outpost of the Highlands.

= 5 km.

31. The Orkneys

For reference see O.S. Routemaster Series, Sheet 1

LB   Loch of Boardhouse
LH   Loch of Hundland
LK   Loch of Kirbister
LS   Loch of Swannay

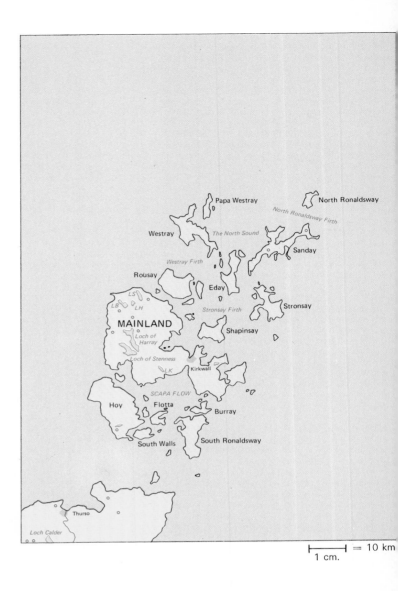

├——┤ = 10 km
1 cm.

The Orkneys are essentially an extension of Caithness and consist of a dissected and submerged platform of old red sandstone. The islands have some of the finest examples of marine erosion in Britain, e.g., the Old Man of Hoy, a stack of old red sandstone rising to 150 m above a lava platform base. Ice from the Scandinavian and Scottish ice-fields deposited variable thicknesses of boulder clay over the islands. This resulted in uniform characteristics with gentle surface gradients, poor drainage and considerable deposits of peat. The one exception of the group is Sanday, the lightest island on the image, which has large areas of depositional sand.

Separated from mainland Scotland by 10 km of the rough Pentland Firth, the Orkneys are linked to Scrabster by a daily ferry service from Stromness. These islands lie in the track of frequent deep meteorological depressions and are therefore battered by gales and virtually treeless.

The economic basis rests on arable farming and 40% of the land is under crops and grass. Stock rearing and fattening are important and dairy farming produces surplus milk that is sent to the mainland for processing or to the Shetlands. Some barley is grown but the crops are mostly oats and turnips. Although the majority of farms are small family holdings they are often highly mechanized and attempts to reclaim heathland are being made.

Kirkwall is the largest town and its industries reflect the dependence of the Orkneys upon sea and farm produce. The major industry is dairy processing for powdered milk, cheese and butter but other industries include a distillery, kipper factory and a small tweed mill.

Perhaps the most famed location in the Orkneys is Scapa Flow, which was an important Royal Naval base during the two World Wars, and an operational base for 20 000 servicemen. It witnessed the surrender of the German High Seas Fleet in 1918 and its subsequent scuttling in 1919. In 1951 the base was closed and this had an adverse effect on the local population.

78

32. The Shetlands

For reference see O.S. Routemaster Series, Sheet 1

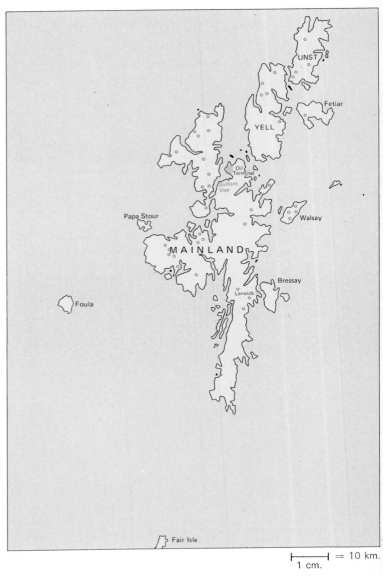

UNST

Fetiar

YELL

Oil
Terminal

*Sullom
Voe*

Papa Stour

Walsay

MAINLAND

Bressay

Lerwick

Foula

Fair Isle

├────┤ = 10 km.
1 cm.

These islands, lying approximately 59°N, are actually closer to Bergen in Norway than to Aberdeen, their principal sea link with the British mainland. The Shetlands consist of over 100 islands or islets, of which 18 are inhabited. Isolation from the mainland and a close proximity to Norway resulted in a long history of Norse colonization and culture, much of which is still present in Shetland activities.

Physically, the Shetlands are composed largely of ice-moulded igneous and metamorphic rocks and are a detached part of the Highlands. The dissected plateau has been submerged to form the penetrating inlets, known as geos, between well defined headlands. Geos are products of marine erosion whilst voes exhibit no evidence of such erosion and they are generally assumed to be the result of subsidence. No part of the Shetlands is more than 5 km from the sea.

Acid soils, low temperatures and a high incidence of fog in summer combined with strong, salt-laden gales restrict farming activities. Arable land is confined to about 4% of the total area with oats, hay and potatoes the principal crops. During recent years, however, large-scale drainage and increased use of mechanization have seen an increase in the number of people involved in farming to the detriment of the fishing industry. Most commercial fishing is centred at Lerwick, the fish being shipped to Aberdeen. Other industries include tourism and hand knitting. Lerwick is the commercial and administrative centre and has approximately 30% of the population.

Conditions have changed somewhat with the opening up of the oilfields in the East Shetlands basin. At Sullom Voe, the site of an old Royal Air Force base 40 km north of Lerwick, and the largest sea loch in the Shetlands, a major oil terminal has been established. This is now the principal North Sea oil terminal. The oil industry has also had an important effect on Sumburgh airport which has been enlarged to enable it to play an important role in the channelling of men and materials to and from the northern rigs.

1 cm.   = 6 km.

I. Exmoor

Green areas represent the moor and heath.

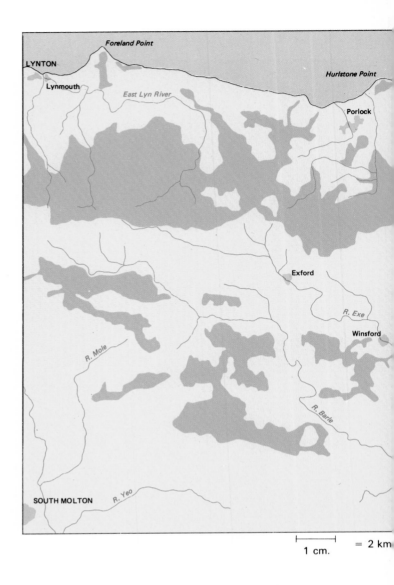

1 cm. = 2 km

This image shows the Exmoor area of SW England with its varied land cover of rolling moorland combined with pastureland and deeply incised wooded valleys. The coastline to the north overlooks the Bristol Channel and contains some of England's highest coastal viewpoints.

This attractive combination of moorland, woodland, pasture and coastline led to the area being designated as Exmoor National Park in 1954. The National Park comprises some 694 square km of land rising to 520 m at its highest point, Dunkery Beacon. About 57% is farmland, 28.5% moorland and 10% woodland, the remaining 4.5% being habitation and roads.

The topography consists of broad summits with level skylines. A series of plateau-like surfaces are dissected by swift flowing streams. Exmoor consists mainly of Devonian sediments of slate, sandstone and shale affected by Armorican folding in Permo-carboniferous times. Softer and younger sediments (Permian, Trias, Lias) are preserved in faulted trenches to the east near Porlock in the south of Exmoor.

The coast of Exmoor is unusual in a long convex slope often extends down almost to sea level before it is eroded to form a marine cliff. West of Lynton these sloping coasts rise to more than 320 m.

The prime task of the Exmoor National Park Authority is to preserve and enhance the natural beauty of the landscape. In order to achieve this the National Park Department have made surveys of vegetation and land use using air photographs and satellite sensor images, with which to monitor change.

The main blocks of moorland can be readily indentified on the satelite image. Simple photographic enlargements of geometrically corrected Landsat MSS images have been examined by the National Park Authority using the Image Digital Processor 3000 in the National Remote Sensing Centre. Classification using all four bands of the Landsat MSS imagery has produced a land-cover map.

The most useful imagery for vegetation mapping it often Band 7 or a false colour composite. Changes in vegetation can be detected by comparing images of different dates. A contrast of Landsat MSS images of 2 July 1977 and 23 April 1982 were processed on the GEMS image processor at Farnborough to reveal the change from moorland to pasture. The changes which ranged in size from about 4 hectares to about 15.5 hectares, have been successfully mapped.

In order to make the fullest use of remote sensing techniques it is normally necessary to extract information from a satellite sensor image such as this in combination with information from air photographs and ground survey.

n.　　= 1 km.

II. The Mendips

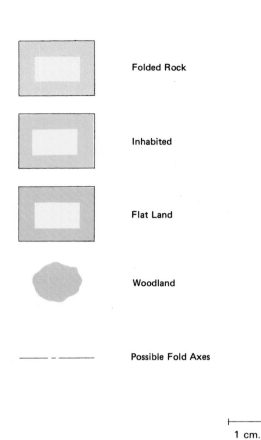

Folded Rock

Inhabited

Flat Land

Woodland

------ - ------ Possible Fold Axes

1 cm. ⊢————⊣ = 2 km.

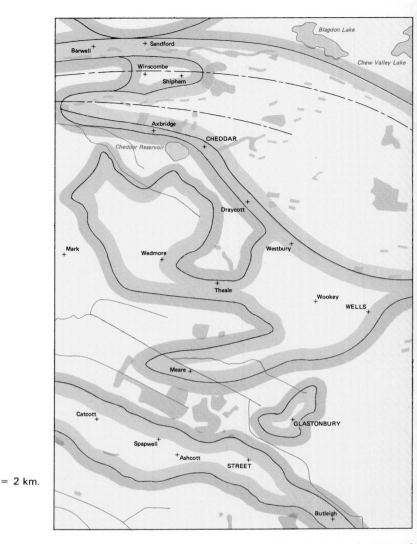

This area demonstrates how colour patterns may be interpreted in terms of rock type and structure even though few rock outcrops may be identified. The inter-relationships between varying geology, topography, soils and drainage often induce patterns in settlements, communications and land use that may then be broadly interpreted. However, interpretation of geological features at a scale as detailed as 1:100,000 in an area dominated by vegetation is not usually possible with Landsat MSS imagery.

Pattern analysis suggests at least three rock units in the area. The most striking pattern is in the northern third of the image which is dominated by the deep magenta patches indicative of forested ground on limestone uplands. The elliptical outlines suggest folded strata, although another interpretation is that of horizontal beds outcropping around elongated hills. Folds suggest an area of sedimentary or metamorphic rocks rather than igneous rocks. Associated with the patches of forest are extensive grasslands which as suggested by the lighter to darker reds in the image are likely to be grazing ground. The absence of either purplish-red linear markings or patches of darker blue on the image indicates terrain lacking large drainage channels and nucleated settlements. The second area is a large peat bog where a different rock type is suggested by areas of grassland traversed by large drainage channels. Many of these channels are so straight as to suggest man-made drainage ditches or rhynes. The visibility of these rhynes at this resolution indicates that they are several metres in width and their necessity suggests poorly drained flat land such as in an area of clay or recent sediments where clays and silts abound. A waterlogged area would explain the absence of sizeable nucleated settlements. This is one of the most productive dairy areas in the UK and is a source of peat for horticultural purposes.

The boundary of the third areal pattern is not always easy to define but is generally differentiated by the dominance of a blue and yellow colour pattern. The darker blue areas denote nucleated settlements. A most conspicuous band of spring line settlements follows and enhances the outline of the folded strata. The lighter blue and yellow patches are indicative of a range of land uses varying from bare rock, for example, quarries, to bare or poorly vegetated soils. The lack of vegetation on some soils may be a function of the time of year that the image was taken and suggests an area of cultivated crops rather than grassland for grazing. The fact that practically all the settlements are associated with this spectral pattern where large drainage channels are not common suggests an area of different rock type and soil above the water-logged area. The large number of small nucleated settlements and the three large lakes or reservoirs suggest gently sloping topography.

Although the area is probably underlain by at least three rock types it is not possible to identify them with any degree of certainty without some previous knowledge of the area. In this respect interested readers may wish to compare the Landsat MSS imagery with the Geological Survey sheets Number 280 (Wells) and 296 (Glastonbury), which include the same area at scales of 1:63,360 or 1:50,000 to realise that only the main geological features may be interpreted from Landsat MSS imagery.

III. The Rhondda

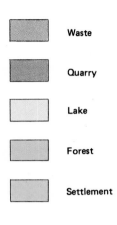

Waste

Quarry

Lake

Forest

Settlement

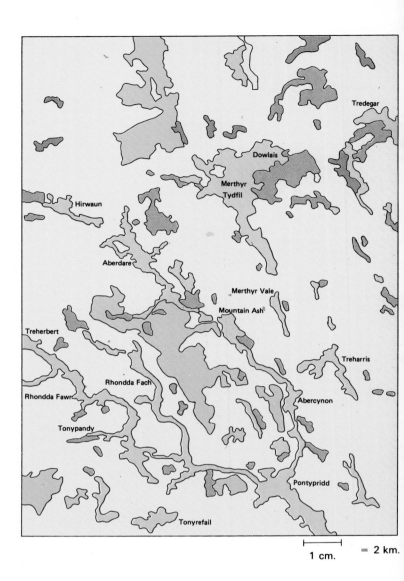

1 cm. ⊢———⊣ = 2 km.

The scene shows the well known South Wales Mining Valleys, from Treherbert in the west to Tredegar in the east, and as far south as Pontypridd.

The linear settlements in the valleys stand out clearly. Mines and houses stand close together, and the transport routes and rivers compete for space in the narrow valley floor. Perhaps less obvious is the sudden halt of the coalfield on a line from Hirwaun (west) to Tredegar (east). At this point the coalfield comes to the surface. North of this line is an outcrop of Carboniferous limestone, and a zone of quarries, apparent on the scene but in appearance similar to the settlements. Beyond this the landscape changes rapidly to that of the reservoirs, forests and moorland of the Brecon Beacons.

Four of the famous valleys can be clearly seen. From west to east they are the Rhondda Fawr, the Rhondda Fach, and then the Valley of the River Cynon (from Aberdare to Abercynon) and the major valley of the River Taff (from Merthyr Tydfil to Pontypridd, and then to Cardiff).

The smaller valleys of Cwm Bargoed and Nant Bargoed are apparent before the Rhymney Valley in the north-east of the scene.

Much of the landscape of the mining district has been disturbed in the past, and many old waste heaps are now covered by vegetation. The image shows more recent waste features on which there is little or no vegetation. The appearance of the waste is similar in infra-red false

colour to the urban settlement features, with the darker blue colour typical of a bare surface. Waste tips with a light growth of vegetation appear a blue/green colour. In the north of the scene are large areas of waste associated with opencast mining. These mines have been very productive economically, but leave a scar on the scenery that is only now being tackled by landscaping and aforestation. Adjacent to the valleys in the centre of the scene are the more typical 'slag-heaps' of the deep mining area. These heaps are typically on the shoulders of the valleys, near to the mines that are on the valley floor. Waste is carried up the slopes by means of aerial hoists, or conveyors, or by vehicle. These heaps are a source of concern not only because of their unsightliness but because some of them are considered to be unstable and could endanger local communities. There is a programme of land improvement now underway.

A general improvement of the landscape is being undertaken by the Forestry Commission, and St Gwynno Forest in the centre of the scene (between Rhondda Fach and Mountain Ash) is typical of large scale changes in land use. Economically, however, the area is depressed and within the settlements there are many derelict of deserted shops, houses and factory units that cannot of course be seen at this resolution of imagery.

= 1 km.

IV. The Fenlands

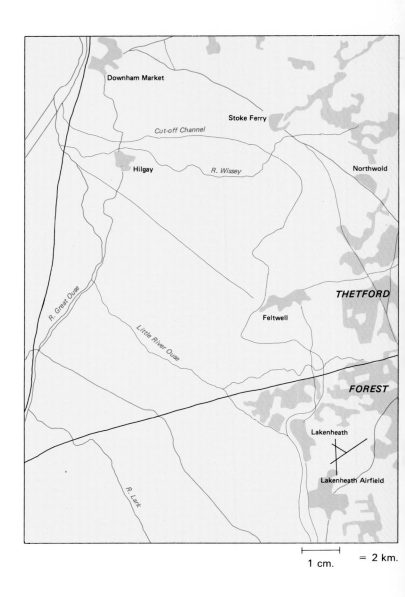

1 cm. = 2 km.

This standard colour composite image covers an agriculturally productive area in East Anglia which includes part of the Fens and the edge of Thetford Forest. The image was recorded on 20 April 1982.

The area has reasonably large fields for the UK but they are still difficult to distinguish because only a small number of pixels occur totally within each field. With the image being taken in April many of the fields are bare soil in which sugar beet, potato and vegetable crops are still to be sown. Bare soil has two distinctive tones on the image, either black or bluish white. The part of Thetford Forest on the eastern edge of the image also appears black. Redness on this image is associated with crop growth. The brightest reds relate to early sown cereals and grassland. Later sown cereals appear light red or mauve.

The contrast between the two soil colours is particularly striking and relates to very different soil types. The soils which appear black are peat soils which have a similar colour of the ground. Light sandy soils and loamy sands near Thetford Forest, and areas of mineral soils within and at the edge of the Fen area, are bluish white. Agricultural drainage and cultivation within the area have resulted in peat wastage by wind erosion so that the depth of peat soil is now quite variable. As wastage progresses the peats become peaty loams, then organic mineral soils and finally mineral soils. Soil colours which are intermediates between black and bluish white therefore exist and indicate these different soil compositions.

The main potential of satellite remote sensing for agriculture in the UK is for crop mapping as a source of early intelligence on levels of crop production and for monitoring changes in agricultural land use. These applications have been slow to develop in Europe in comparison with the USA for two reasons. Firstly, the inadequacy of the 80 m resolution of Landsat MSS images in relation to relatively small field sizes; and secondly, the unreliability of temporal coverage due to cloudy weather conditions. However, early assessment of 30 m resolution Landsat Thematic Mapper and 20 m resolution SPOT simulation images indicates a marked improvement in the definition of individual fields which will greatly increase the potential for agricultural uses in the UK and Europe. In addition to improvements in resolution the off-nadir viewing capabilities of SPOT HRV are expected to improve significantly the availability of multi-temporal data.

= 1 km.

V. Hadrian's Wall

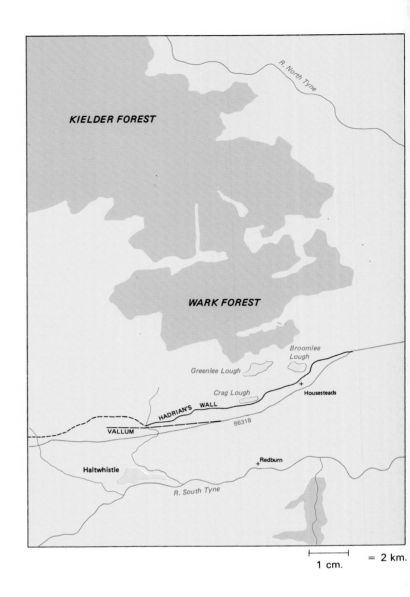

1 cm. = 2 km.

Landsat 1 and 2 MSS imagery, with its relatively coarse resolution, is not ideally suited to the sensing of small archaeological features which conform to the landscape. When a feature occurs over a relatively large area, particularly when it coincides with easily identifiable landscape elements, it may be possible to trace its position and extent on a Landsat MSS image.

Hadrian's Wall is such an example, running approximately 130 km between Wallsend, on the east coast, to Solway Firth, on the west, and was one of the strongest and most elaborate of all Roman frontier fortifications. It was more than a mere wall, consisting of a ditch and a wall, behind which was a road called the Military Way and an elaborate set of earthworks referred to as the Vallum. The purpose of this latter feature is unknown.

By increasing, or stretching, the contrast of the image and utilizing the shadows of the landscape cast by oblique illumination, it is possible to exaggerate the natural variations of the terrain. This in turn helps identify the topography along which the wall runs. Running across the image may be traced the course of the B6318, often referred to as the Military Road. This was built following the 1745 uprising when Bonny Prince Charlie invaded England by way of the west coast, to the annoyance of General Wade who was waiting at Newcastle. Unfortunately for latter day archaeologists, no east/west road existed that was suitable for artillery so following the defeat of the Scots at Culloden the Government built a road, the foundations of which consisted of the Roman Wall.

Just inside the eastern edge of the image, the road veers away from the wall which traverses the hilly, inhospitable region of the Whin Sill, an extensive doleritic extrusion. The wall clings to the top of this very prominent feature which can be easily picked out on the image as an area of shadow running across virtually the remainder of scene. Sewingshield Crags may be identified overlooking Broomlee and Greenlee Loughs. Here the wall was built on top of the crags with a sheer drop of 60 m on the northern face.

Further along the wall it is possible to locate the position of Housesteads, one of the most important forts on the wall covering a site of approximately 5 acres. This has been described as one of the finest Roman forts to be found in Europe.

To the west of Housesteads the wall follows the line of Holbank Crags and Steel Rigg. Here its course runs along the southern edge of the darker area formed by the shadows cast by the Whin Sill. To the south of the wall is a lighter area that ends along a line running in an approximate east/west direction. This line coincides very closely with the position of the Vallum.

⊢———⊣ = 1 km.

VI. The Tay Estuary

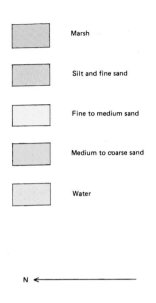

Marsh

Silt and fine sand

Fine to medium sand

Medium to coarse sand

Water

N ←

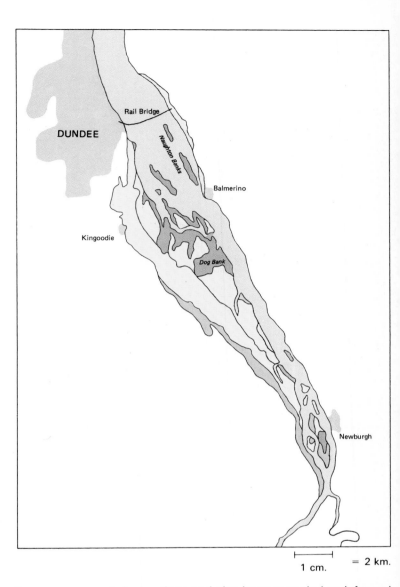

DUNDEE

Rail Bridge

Naughton Banks

Balmerino

Kingoodie

Dog Bank

Newburgh

1 cm. = 2 km.

The accompanying image shows a false colour composite of the upper Tay Estuary with the state of the tide at near low water. The image is produced from Landsat MSS Bands 5, 6 and 7 and the scene was recorded on 24 October 1976.

The objective of this image is to consider the water area in the image which is shown between Perth in the west and Dundee in the east. A consideration of the image would, however, be incomplete without considering the surrounding countryside on both banks of the estuary. Because the overpass of the satellite occurred near low water on a spring tide the extensive tidal flats of the upper Tay Estuary are clearly visible.

Along the boundary between the land surface and the tidal flats are broad fringes of orange-pink tonality in the west (lower) part of the image. These fringes typify reed beds and are particularly clear around the grasslands of Mugdrum Island off Newburgh. The reed beds occupy extensive areas eastwards along the northern (left bank) shore towards Kingoodie.

A distinct interface exists between the water filling the estuarine channels and creeks and areas exposed above water. The tonal variation on the tidal flats from dark greys immediately off the reed beds, through lighter greys on the outer flats, to pale blues besides the principal channels and creeks, and the outermost parts of Dog Bank result from differing water retention by the sediments. The pattern of sediments varies from silts near the marsh margin to fine and medium sands across the tidal flats towards the channels. This pattern has been seen on

ground observations and also been recorded on infra-red aerial photography such that they were able to confirm the results recorded on this Landsat MSS image.

The distribution of channel margin sandbanks from Balmerino eastwards follows the general patterns that were established when the upper estuary was charted. However, the linear Naughton Banks east of Balmerino Point have changed substantially in position since the survey 4 years before the imagery was collected.

Before any quantitative comparison could be carried out between the Landsat MSS image and existing charts which ideally record the features being interpreted in the estuary, the image had to be geometrically 'rectified'. Ground control points were selected from the Ordnance Survey map sheet of the area that were visible on Band 7 of the Landsat MSS image. Once the rectification had been completed, comparison with a survey of the upper estuary carried out by the Tay Estuary Research Centre in 1972–73 could take place. Over most of the upper estuary the differences between the low-water mark on the chart and the low-water mark on the image were within the permissible error for the survey or about one or two pixel edge lengths (79 m) which for Landsat MSS is about 150–200 m. However, the differences for the Naughton Banks between Balmerino Point and the rail bridge are substantially larger than the errors involved in the analysis and represent real movement of these banks between 1972–73 and 1976. With the availability of repeated imagery the sandbanks could be monitored without the need for additional ground data and aerial photography.

1 cm. = 1 km.

VII. Elgin

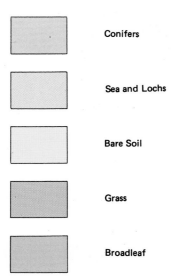

Conifers

Sea and Lochs

Bare Soil

Grass

Broadleaf

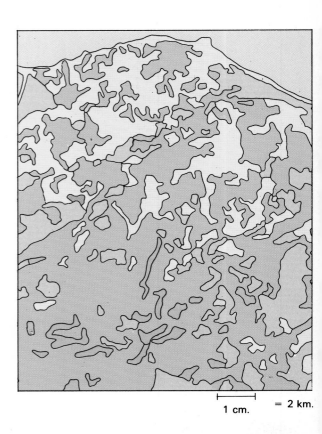

1 cm. = 2 km.

The Elgin Project, a joint project with Grampian Regional Council and the Forestry Commission, was set up to assess the potential of land cover classification of Landsat MSS imagery as an aid to regional planning and forest cover monitoring. The project site, 25 km by 25 km around Elgin, takes in the flat coastal plain and the foothills of the Grampians to the south. Forestry occurs in both main site types, as larger conifer plantations and smaller mixed woodlands in an agricultural matrix in the lowlands and as predominantly coniferous plantation and moorland in the uplands. Commission records also indicated significant recent afforestation in the uplands, providing a test of how quickly cover changes could be identified from satellite data.

Image analysis was carried out on the GEMS image analysis equipment at the National Remote Sensing Centre at RAE, Farnborough. Recent Landsat MSS imagery was limited, Landsat 3 line start problems losing the western third of the data for one of the satellite paths covering Elgin. The Landsat 2 MSS scene 221/20 of 20 April 1976 was the most recent to give good visual discrimination between woodland and agriculture and an extract covering the project area was geometrically rectified to map co-ordinates and resampled to 50 m pixels. An automatic two step linear stretch was applied to all bands to enhance cover type differences and a colour composite was produced using Bands 4, 5 and 7.

Classification was based on known forest cover 'training areas' derived from Commission survey maps, aerial photography and ground checks. The spectral ranges of all four Landsat MSS Bands were sampled in the training areas, using a box classifier, and extended to like areas in the rest of the image. Other cover types were classified in the same way. The following categories were recognized:

1. Coniferous forest.
2. Deciduous and scrub woodland.
3. Agricultural cropland and grassland.
4. Bare soil and newly seeded land.
5. Wetland and shallow marginal water.
6. Deep water, including sea.
7. Unclassified.

The unclassified areas in the lowlands were mainly urban land, coastal dunes and felled woodland. In the uplands, unclassified equated to moorland.

The total woodland classification, conifer plus broadleaf, is shown opposite. The coastal Roseisle and Lossie blocks show clearly. The division between lowland and upland runs from south of Monaughty (centre west) to the north end of Teindland (lower east).

Similar classification overlays at 1:50,000 were compared to O.S. maps to determine classification accuracy. The results were encouraging in the lowlands, 94% of sample points classed as woodland being correctly identified. Mapped woodland which was shown as agriculture or unclassified on the imagery (23% of samples) was mainly known clear felling—the square gap in the centre of Monaughty is an example. The upland results were far less successful, partly due to confusion between woodland/moorland elements and partly as afforestation was less than five years old. It appears that new planting has to close canopy (over eight years of age) before a woodland reflectance is apparent. Work elsewhere suggests that autumn imagery is more suited to woodland/heathland discrimination. Work continues on the project.

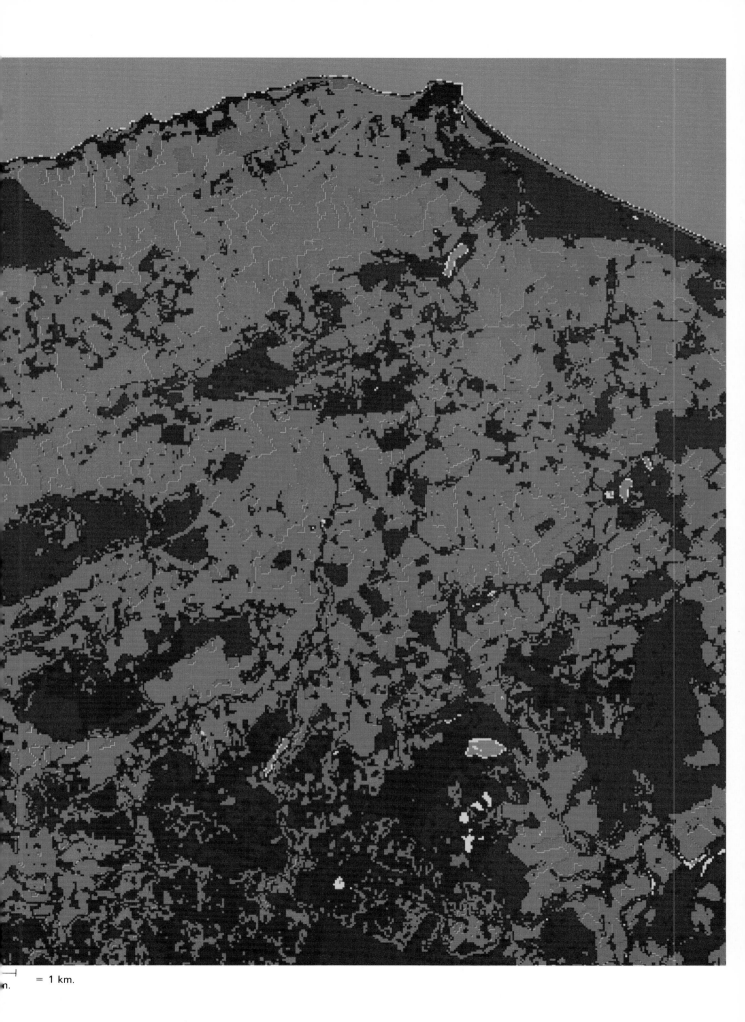

= 1 km.

95

VIII. Lewis

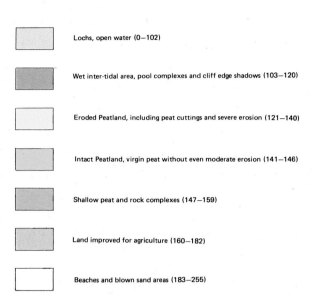

Lochs, open water (0—102)

Wet inter-tidal area, pool complexes and cliff edge shadows (103—120)

Eroded Peatland, including peat cuttings and severe erosion (121—140)

Intact Peatland, virgin peat without even moderate erosion (141—146)

Shallow peat and rock complexes (147—159)

Land improved for agriculture (160—182)

Beaches and blown sand areas (183—255)

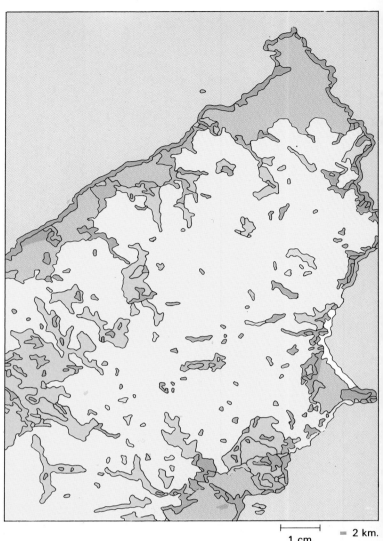

1 cm. $\longmapsto$ = 2 km.

The accompanying Landsat MSS image of the northern part of the Isle of Lewis shows an analysis to classify improved peatland and other land cover within the scene. There are seven classes of land cover which are colour coded and ranged on a grey scale 0 to 255. By providing these grey scales it is then possible to repeat the analysis in an adjoining area and in another time provided the illumination remains the same.

The image has been produced by adding the original four Landsat MSS Band variables to the ratio of Bands 5 to 7 and Bands 4 to 6.

A remote sensing research programme for peat resource surveys was initiated at the Macaulay Institute for Soil Research in 1975. Since then it has been demonstrated that a multi-level and multi-spectral approach is the most economic method for mapping large areas. To fully assess terrain characteristics over a wide area, including all peatland categories in particular, a carefully planned stratified random sampling strategy based on a hierarchy of scales is recommended. Ideally the sampling blocks as a whole should contain all the desired mapping units required for controlling the final classification.

The research programme was initiated to assist in the routine mapping and assessment of Scottish peatlands, particularly in remote regions where large area ground surveys have proved costly. The primary aim of the research was therefore to establish a database of peat and terrain information of value to potential peatland development.

The production of a medium scale 1:100,000 map illustrating the peat and terrain categories of Lewis was the final objective of the programme.

The supervised classification techniques used in this programme were based on sample photogrammetric plots of test areas. The study showed that a Landsat image with the appropriate processing can produce up to 80% of the information required to map the peat and terrain categories at the required scale.

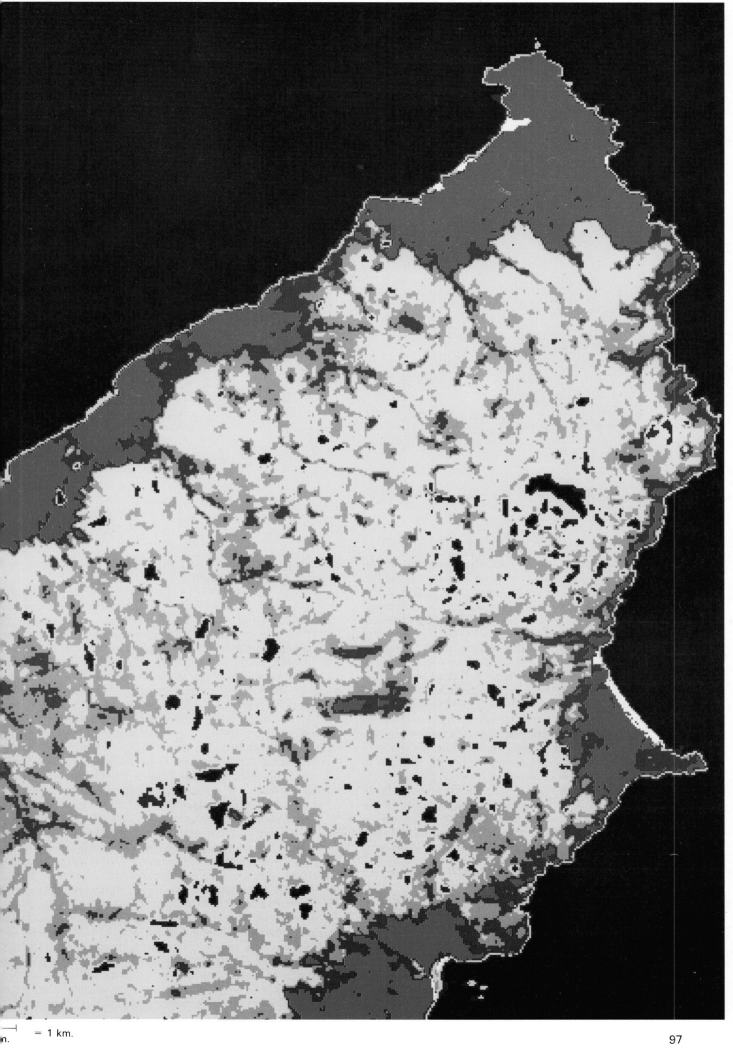

= 1 km.

A. TIROS-N AVHRR

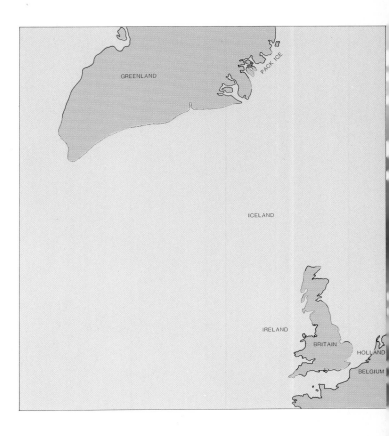

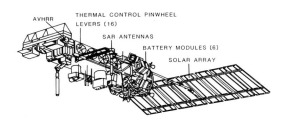

Figure 14. The NOAA Tiros-N satellite.

The TIROS-N and NOAA 6–8 satellites operate with an Advanced Very High Resolution Radiometer (AVHRR). The satellites in this series are near polar Sun-synchronous operating with a four-channel scanning radiometer which is sensitive to red, near infrared and thermal infrared radiation. The channels have been selected to provide improved determination of hydrographic, oceanographic and meteorological information with the aid of the multispectral capability. Although the resolution is low (see Table 1 below), there is increased interest in the satellite imagery for land resources especially as an inventory of vegetation over large areas.

The visible and near infrared channels can be used to discern clouds, the land-sea interface, the presence of snow and ice and by comparing sequential imagery to establish variations in coverage of snow or spread of ice, especially as icebergs. With the thermal infrared channel it is possible to determine the surface temperature of the sea. This series of satellites is expected to provide a global observation service from 1978 up until 1989. The AVHRR is capable of providing data in both the visible and the infrared channels during the day and in the infrared channels during the night.

This TIROS-N AVHRR image was received on 27 June 1979 at 15.23 hrs GMT. The image shows a very well formed low pressure cyclone south-east of Greenland. The cloud system can be seen to be drawn into the cyclone. In the top of the image can be seen the snow-covered mass of Greenland with light cloud cover and adjoining the well-defined north-eastern coast line. The pack ice showing up as a pink/grey tone can be seen off the north-east coast of Greenland. Iceland, Ireland and Norway are cloud-covered and their position has been indicated on the map. Those coastlines that are partly visible are indicated by a broken line. The outline of Britain can be seen except for the area on the eastern and north-western coasts covered by cloud. The Faeroe Islands are faintly visible in the clear 'corridor' between the cloud fronts. The coastline of northern France, Belgium and part of the Netherlands including the Zuider Zee can be clearly seen. The Channel Islands can be seen off the French coast. The lines in the clouds west of Britain are most probably caused by high flying jets. The small scale of this image and the large amount of cloud covering the land masses prevents the vegetation variations throughout Northern Europe from being identified. Because the image is produced from the visisble and near infrared bands the temperature differences in the sea cannot be detected and the sea where not covered by cloud is a uniform shade of blue. As the swathwidth is 3,000 km it is possible to record this large area in one scene. The importance of the TIROS-N satellite to record future weather patterns in Britain, which are determined by activity in the Atlantic, is well illustrated in this image.

Altitude	833–870 km
Number of radiometers	4
Spectral ranges	0.55–0.68 μm
	0.725–1.10 μm
	3.55–3.93 μm
	10.5–11.50 μm
Swathwidth	3000 km
Resolution	1.1 km

Table 1. TIROS-N and NOAA AVHRR details

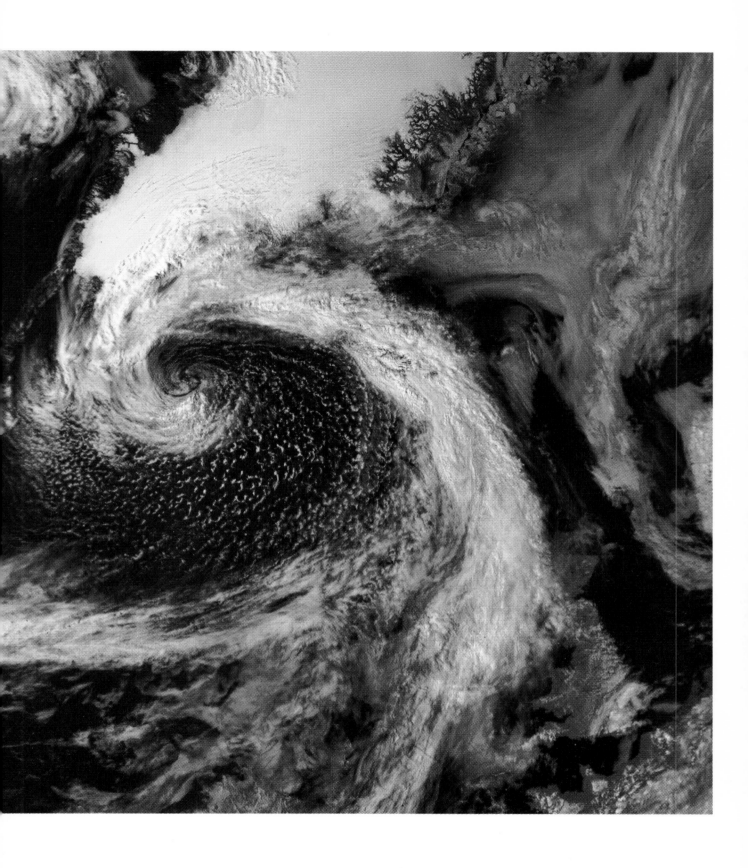

B. Landsat Return Beam Vidicon (RBV)

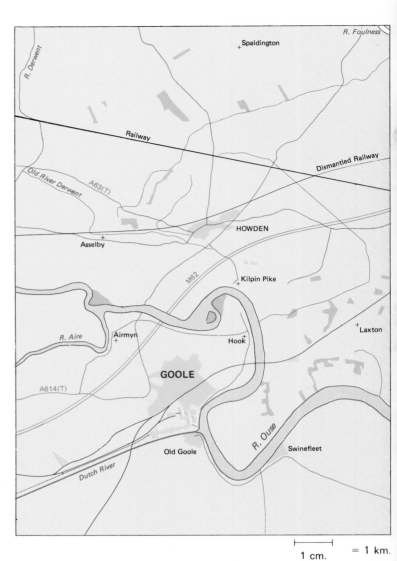

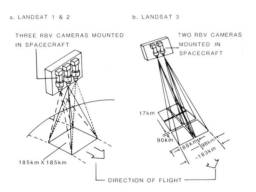

a. LANDSAT 1 & 2

THREE RBV CAMERAS MOUNTED
IN SPACECRAFT

b. LANDSAT 3

TWO RBV CAMERAS
MOUNTED IN
SPACECRAFT

185km X 185km

DIRECTION OF FLIGHT

Figure 15. RBV scanning patterns.

1 cm. = 1 km.

The RBV is one of two data collecting devices on board the Landsat 1–3 satellites, the other being the MSS. The RBV systems differ in that Landsats 1 and 2 contained a multichannel 3-band RBV system while Landsat 3 contained two single cameras. Figure 15(a) shows the three cameras for Landsat 1 and 2 and Figure 15(b) the two cameras on Landsat 3 together with swathwidths. A vidicon camera enables high resolution images in the 0·35–1·1μm region of the electromagnetic spectrum (EMS) to be recorded with the aid of an electronic charge coupled device. With the Landsat satellite the image is recorded on an RBV tube which provides greater sensitivity at low light levels and higher resolution than the ordinary vidicon tube. Vidicon cameras can produce realtime and near real-time imagery of the Earth's surface; applications in weather satellites are now familiar to those watching weather forecasts.

The imagery from the RBV cameras on board the Landsat 1 and 2 satellites has a resolution of 79 m, the same resolution as the MSS imagery; this resulted in the

RBV imagery being overshadowed by the spectacular MSS imagery. This preference for MSS imagery has resulted in very little RBV imagery from the Landsat 1 and 2 satellites being available. However, with Landsat 3, because of an increase of resolution to 24 m with the RBV imagery the whole dimensional concept has been improved. The image on the facing page shows the Landsat 3 RBV scene of the Goole area which, when compared with the O.S. Landranger 1:50,000 Series, Sheet Nos 105 or 106, shows the higher spectral quality of this imagery when compared with the MSS.

Incorporated into the RBV systems is a réseau which is imprinted onto the image providing a reference system from which measurements to objects on the image can be taken and by which the image can be rectified. The réseau consists of a number of accurately spaced crosses which are recorded on the image.

As stated previously, the accompanying image and map show how RBV imagery from Landsat 3 can be used both for map construction and where maps already exist for map revision. The single band panchromatic image may be a limiting factor when attempting to discriminate vegetation types and the land/water interface. The two islands in the River Ouse cannot be clearly seen in the image because the woods and water both appear light grey. The straight line of the railway in the top half of the scene shows how well the image has been rectified. The M62 motorway shows up clearly except for interchanges 36 and 37. Difficulty would be experienced with this single-band image to differentiate the road network in Goole, while the field boundaries can be clearly delineated.

Characteristics of RBV	Landsat 1 and 2	Landsat 3
Number of cameras	3 (covering same area)	2 (covering adjoining areas)
Spectral bands	0.475–0.575 μm 0.580–0.680 μm 0.698–0.830 μm	0.505–0.750 μm
Ground area	185 × 185 km	99 × 99 km
Resolution	79 × 79 m	24 × 24 m

Table 2. RBV systems on board Landsats 1–3.

= 0.5 km.

C. Seasat Synthetic Aperture Radar (SAR)

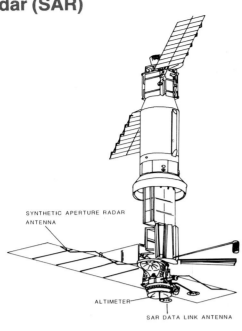

Figure 16. The Seasat satellite.

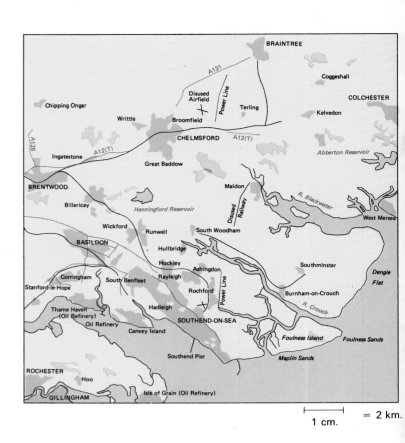

1 cm. = 2 km.

The Seasat satellite was the first to incorporate a high resolution Synthetic Aperture Radar (SAR). Synthetic refers to the use of a short onboard antenna on the satellite and using the distance travelled in the time interval between the transmission and the receiving of signal as the 'antenna'. The satellite was launched on 26 June 1978 and operated for 106 days until 9 October when it failed because of a massive short circuit in its onboard electrical system. The satellite was primarily designed for recording data from the sea surface but results have shown that its application for recording land surface features may be of even greater importance. The image on the facing page gives a good indication of the SAR on board a satellite and its potential. in the three-month life of the satellite the amount of data collected was such that a surface area of 126 million square kilometres was covered.

Altitude	800 km
Wavelength	0.235 m
Swathwidth	100 km
Resolution	25 × 25 m
Look angle from nadir	20.5°
Total beamwidth	6.2°

Table 3. Details of the SAR on board the Seasat satellite.

The radar system is active as compared with cameras and scanners which are passive and rely on reflected signals from an external source, for example the Sun, to provide the energy. The radar emits a series of high-powered signals which are returned to the sensor at strengths dependent on the angle of reflectance of the signal and with the roughness and didective contrast of the surface object. Because of the power being supplied by the satellite and because of the wavelength of the sensor

where transmission is 100% in the 'L' Band (Figure 16), the images can be obtained in all weather conditions and by day or night. The image that is produced by radar is not a 'normal' image, as the characteristics of the image are dependent on the look angle of the sensor and the direction or orbit of the satellite. If the path of a satellite is parallel to a mountain range the signal from the 'front' of the mountain will be strong and there will only be a weak or even no return signal from the other side of the mountain. The resultant image will be a bright (strongly reflected) 'front' of the mountain with a dark 'back' giving the impression of a shadow and making the mountain appear to stand out. If the path of the satellite crossed the mountain range there would be very much less contrast and no 'shadow' would be formed, and therefore the mountain would not be recognized from the image.

The image on the facing page of Essex, Kent and the Thames and Blackwater estuaries is an area of relatively flat ground and 'shadows' produced form high features are not present in large quantities as would occur in mountainous areas. The image was recorded on 16 August 1978 at 6.40hrs GMT and the state of the tide at the time of collecting the data was halfway between low and high tide, on the flow. The path of the satellite is parallel to the left hand edge of the image. The effect of the path on this image is to cause all objects on the surfce, particularly linear features, to be more clearly recorded when they are parallel or running from top to bottom of the page. The Southend Pier and the outfall at Southend are examples of linear features that stand out very clearly. Numerous high voltage power lines (running top to bottom) can be seen clearly and the pylons appear as white dots on the image. there are 24 cables on each high voltage power line strung between the pylons and they, together with the spreaders, act as strong reflectors.

= 1 km.

D. Landsat Thematic Mapper (TM)

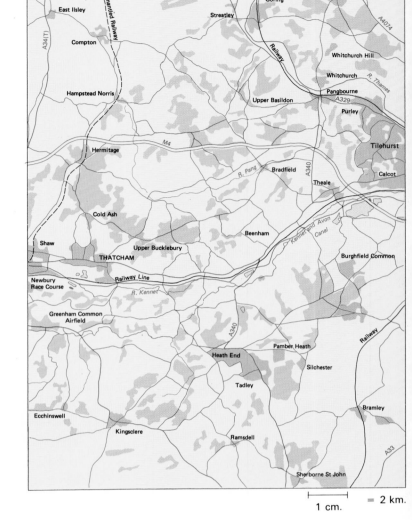

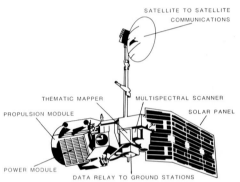

Figure 17. The Landsat 4 satellite.

The thematic mapper is one of two sensors on board the Landsat 4 and 5 satellites, the other being an MSS, the same as that on Landsats 1, 2 and 3. The TM provides a 7-band improved resolution scanner at a lower altitude than from the previous Landsat satellites. The additional bands in the TM are those in the blue and mid-infra-red regions of the EMS with the thermal infrared band being similar to one on board Landsat 3.

Characteristic		Specification
Altitude		705 km
Spectral range	Band 1	0·45–0·52 µm
	Band 2	0·52–0·60 µm
	Band 3	0·63–0·69 µm
	Band 4	0·76–0·90 µm
	Band 5	1·55–1·75 µm
	Band 6	10·4–12·5 µm
	Band 7	2·08–2·35 µm
Ground area		185 × 185 km
Resolution	Bands 1–5 and 7	30 m
	Band 6	120 m

Table 4. Details of the thematic mapper on board Landsat 4 and 5 satellites.

The satellite follows a Sun-synchronous repetitive orbit crossing the equator on its descending mode (going southwards) at about 09.45 hrs LST and the inclination of the satellite is 98.2° to the equator, with the orbit taking 98.9 minutes and a full coverage possible in 16 days. The altitude has been reduced from previous Landsat satellites but the surface area has remained the same; the time

for complete coverage is reduced by two days.

The seven bands for the TM were designed for the following purposes:

Band 1. Coastal water mapping; soil/vegetation differentiation; deciduous/coniferous differentiation.

Band 2. Green reflectance by healthy vegetation.

Band 3. Chlorophyll absorption for plant species differentiation.

Band 4. Biomass surveys; water body delineation.

Band 5. Vegetation moisture measurement; snow/cloud differentiation.

Band 6. Plant heat stress management; other thermal mapping.

Band 7. Hydrothermal mapping.

The TM image on the facing page covers a part of the Kennet Valley which bisects the northern and southern parts of the image. The M4 motorway and interchange 12 are clearly seen north of the River Kennet and cutting across the north-east corner is the River Thames. The outskirts of Reading on the right bank of the River Thames are clear but probably not sufficient to prepare the road network for a map at 1:50,000 scale. The runway of RAF Greenham Common south of Thatcham is clearly visible. The field boundaries and woods are almost too numerous to map, certainly at the 1:200,000 scale of the map above.

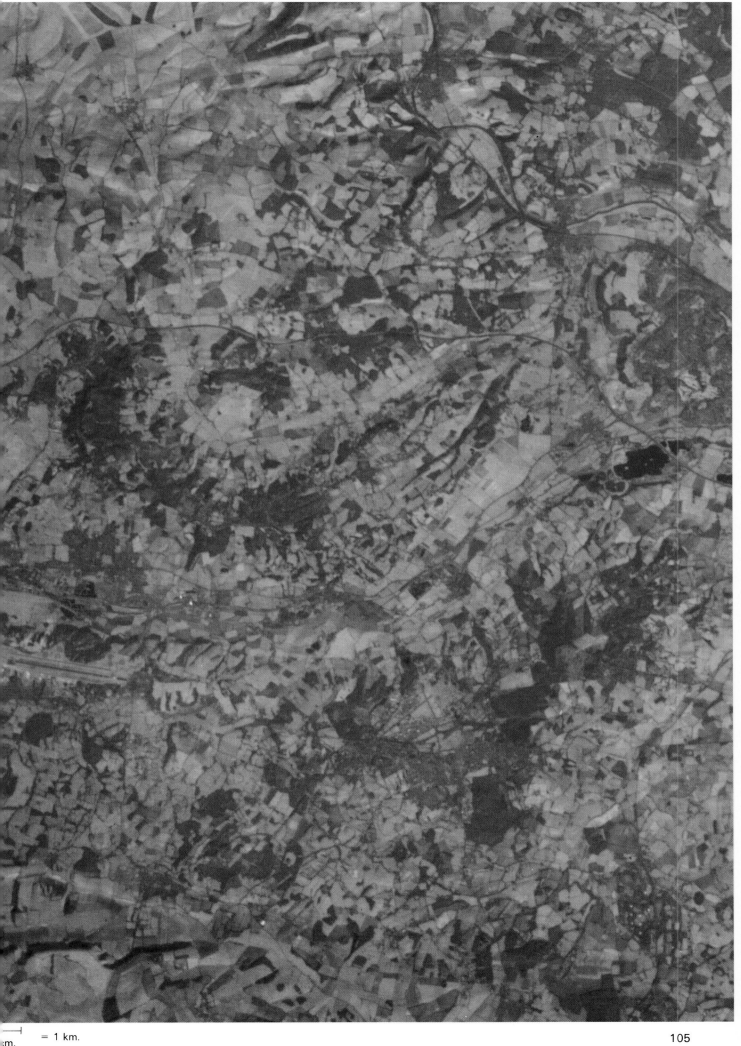

E. SPOT High Resolution Visible (HRV)

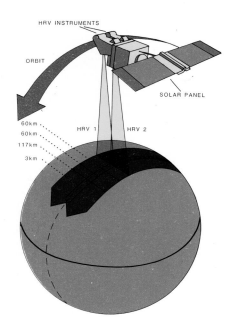

Figure 18. The SPOT satellite at nadir viewing.

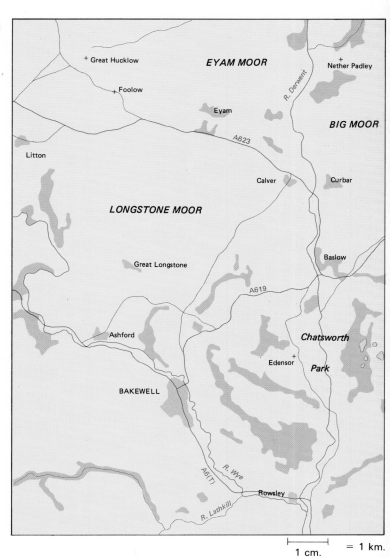

1 cm. |———| = 1 km.

At the time of writing it is planned that SPOT will be the first French satellite to be launched on an Ariane rocket in 1985. The name SPOT is the acronym of Systeme Probatoire d'Observation de la Terre, and a consortium of organisations including the French Centre National d'Etudes Spatiales (CNES) together with European partners (Belgium and Sweden) is responsible for the construction of the satellite.

The image on the facing page was acquired by an airborne multi-spectral scanner in a campaign to simulate the characteristics of the planned satellite data.

The resolution of the SPOT sensor was designed for the small scale subdivisions of much of the agricultural land in many parts of the world and for cartographic as well as many other applications.

The SPOT satellite has a payload of two identical high resolution visible (HRV) imaging instruments together with support equipment of two magnetic tape recorders and a telemetry transmitter.

The HRV sensors can operate in either the multispectral or panchromatic modes. The resolution is higher in the panchromatic mode, as can be seen in the table. Data is collected by an array with 6000 detectors which enables the complete line of the ground scene to be covered in one 'sweep', this has led to this type of instrument being called a 'push broom' scanner.

The orbit of the satellite is designed to be Sun-synchronous, near polar, with an inclination of 98·7° at an altitude of 832 km. The imagery will be collected in the descending mode with the satellite crossing the equator at 10.30 hrs LST, repeating its ground track every 26 days.

The sensor is passive and therefore the data are obtained with aid of reflected radiation. The near infrared radiation enters the HRV through a plane mirror that can be steered by ground control enabling off-nadir viewing. The mirror can be rotated 27° either side of nadir in 45 steps of 0·6°, allowing two strips of 60 km either side of the ground track to be recorded in a 950 km strip.

On this simulated image, Bakewell is seen in the centre lower quarter with the River Wye passing through the town. Downstream the Wye joins up with the River Derwent. The woods, improved grassland and moors can be identified, as well as Chatsworth Park including Chatsworth House and outbuildings.

The reservoir on Middleton Moor and the Emperor Lake at Chatsworth Park are visible. Large networks of roads and tracks are on the image but only the main roads have been recorded on the reduced scale map.

Characteristics of HRV sensor	Multispectral mode	Panchromatic mode
Spectral bands	0·50–0·59 µm 0·61–0·68 µm 0·79–0·89 µm	0·51–0·73 µm
Instant field of view	4·13°	4·13°
Ground sampling interval (nadir)	20 × 20 m	10 × 10 m
Number of pixels per line	3000	6000
Ground swath (nadir)	60 km	60 km

Table 5. Details of the HRV sensor on board the proposed satellite SPOT.

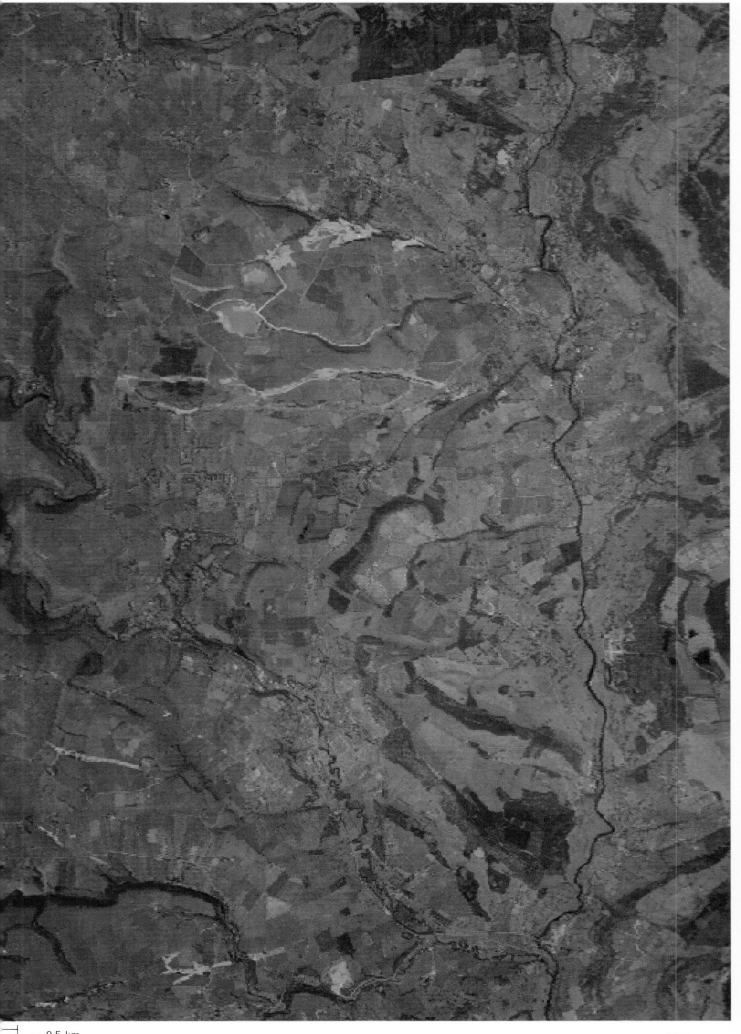

107

Geographic information systems

The terms geographic information system (GIS) or land information system (LIS) are used to denote a system for collecting, checking, integrating and analysing information related to the surface of the Earth.

The sensors on board the Landsat satellite produce large amounts of spatial data; as stated previously the MSS produces about 30 million pieces of information per scene. To be able to handle all this information, and in a sense thereby to justify the collection of such a large amount of data, an efficient processing system is required to convert the data into a usable format. The GIS is one method by which the data can be used. To date there is a surplus of data being obtained by satellite sensors, which are at present not being used to provide all the information which they are capable of. This situation questions whether the data collecting systems are at present beyond the handling capabilities of the operators or beyond the users' requirements.

The resolution of some of the satellite data is such that its application in GIS is, however, rather limited; in the second generation satellites the sensors will provide much more relevant data. The GIS does not at present use remote sensing data to any great extent, the major source of information to date is derived by other techniques which are mainly ground based.

Information for a GIS may be either in a digital or analog format, most of the information is likely to be obtained in analog format from maps. Some of the sources of information are shown in Figure 19. With all this information, often stored at separate centres, but ideally linked to a central information centre, a number of uses or applications for the GIS become apparent: land development; town and country planning; selecting routes for pipelines; routes for new roads, railways or canals; and the registration of interests in land are some of the more obvious examples of GIS.

Data structure

In a GIS the information may be stored and then processed in point, line or area format. All features on the surface of the Earth can be recorded in one of these three formats. Points can be represented by their ground coordinate values x and y. Lines can be represented by coordinates at each end while areas can be thought of as polygons with a number of lines that both enclose and fit the irregularities of an area.

Figure 20 shows the two methods of storing data, in vector or raster format. The map can be translated or digitized into a series of lines or points and these are known as vector format. Satellite-borne sensors like the Landsat MSS sensor scan the Earth's surface in narrow strips and this produces information in raster format. Whichever format is used it must be noted that the resultant positional accuracies of the data finally presented by a GIS cannot be greater than the original data, be it

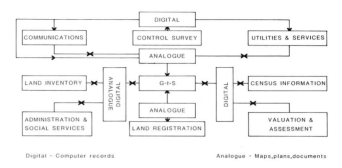

Digital - Computer records   Analogue - Maps,plans,documents

Figure 19. The information flow within GIS. Bullard, 1981, in *Matching Remote Sensing Technologies and Their Applications,* (Remote Sensing Society, Reading, UK).

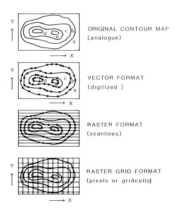

Figure 20. Vector and raster data structures.

the digitized map or the scanned image from a satellite sensor.

In most countries where GIS operates, the major activity at present is concentrated in the data input stage. The creation in the past of maps and plans by national mapping agencies has taken many years and to convert this analog information into a digitized format is both expensive and time-consuming. By the time the bulk of the analog data has been digitized and made available for GIS, the remotely sensed images in raster format will, especially as the resolution improves, provide data for updating and monitoring much of the land information.

Figure 17 shows the way that Landsat imagery could be incorporated into a GIS.

With the recent report in the UK of the House of Lords' Select Committee on Science and Technology, *Remote Sensing and Digital Mapping*, it is anticipated that the two activities will come closer together. It is also anticipated that remote sensing with the aid of digitized maps will provide more information with which to monitor the Earth and its resources. As the resolution of remotely sensed images improves and the GIS operators complete their major data input stage then many more applications will arise for these complementary data collection and storage facilities.

Conclusions and the way ahead

This Atlas has attempted to indicate how remote sensing can be used to detect and monitor activities on the surface of the Earth with particular reference to those occurring in Britain. It has shown how the agencies in Britain, the National Remote Sensing Centre and the Ordnance Survey, have an important role to play in promoting the applications of remote sensing and the production and revision of the National Map Series respectively.

The section on Geographic information systems shows how remote sensing data can be used in the future and indicates how data bases can be brought together to both simplify their access to the public and to integrate the data so that it is more relevant and useful to the user community.

With the improving resolution of the imagery provided by new satellites, the reliability of ground control will have to be correspondingly upgraded and therefore rely more heavily on the O.S. plans where they exist at the required scales. When plans do not exist or where the control is required at a standard higher than that scaled from a plan, ground surveys to fix position will have to be undertaken. For the provision of a GIS there will be a need to consider the more accurate determination of property boundaries, that is better than the 'general boundary' system which exists at present relying on the graphical position as determined by O.S. maps and plans.

The section covering the other satellites and sensors gives an indication of the wide selection of imagery that can be made available for a variety of specialist activities. Of the five specialist images, that of the SPOT HRV is only a simulation of what the image is expected to look like. As the SPOT satellite is expected to be providing imagery later this year, 1985, it would be appropriate when looking into the future to start with and elaborate further with this system before looking into more distant systems.

SPOT HRV

Figure 21 shows the ground stations and satellite ground track for SPOT. The SPOT mission centre at Toulouse is responsible for the control station and for giving priority to the requests from users via the foreign and local stations for the operation of the satellite. Figure 18 shows how the

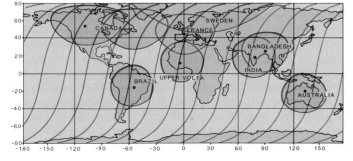

Figure 21. The SPOT data reception stations.

off-nadir capability produces the need to determine priorities, raising the question—should the mode of operation be 10 or 20 m and which pointing satisfies the most users? The SPOT unit will be responsible through the CRIS (Centre de Rectification des Images Spatiales), a space image rectification centre, to process the raw images to different levels or stages of correction.

The standard products as well as custom-made products will be distributed to the users via SPOT Image or the foreign stations. in Britain SPOT data will be obtained from either Toulouse or Kiruna (Sweden) ground receiving stations and distributed through the NRSC.

ERS-1

Of importance to the European Community will be ERS-1 (ESA Remote Sensing Satellite-1) which will be the first remote sensing satellite of the European Space Agency (ESA). The satellite is due to be launched in 1987 and it is being constructed by a number of European companies including Marconi (UK), Dornier (FR Germany), and Matra (France).

ERS-1 will carry a synthetic aperture side-looking imaging radar together with radar scatterometer devices to measure wind speed and wave height, a high precision altimeter and a sea surface temperature measuring instrument. With a minimum swathwidth of 80 km the SAR can be used at two resolutions, 100 × 100 m or 30 × 30 m.

The commercial use of ERS-1 will primarily be in weather, ocean and ice forecasting activities in the near and middle future. The satellite will also be used for marine pollution monitoring, especially to detect and track oil slicks and other harmful products. Despite the ocean and weather applications it is expected that ERS-1 will be used for determining surface structures (geology), surface roughness (mapping) and for the signature of certain geological materials and to be able to establish vegetation conditions. The example of the Seasat image given in this Atlas shows the capabilities of an SAR based on a satellite both for its land and sea applications.

The ERS-1's capability to generate all-weather high resolution images over the land with the SAR can be seen to complement other satellites operating in the optical range, for example Landsat 4 and 5 and SPOT HRV when it becomes available.

Future missions in the ERS series will include a land application satellite AERS-1, which is already being developed.

Future satellites being developed in the USA, Canada, India and Japan will further broaden the range of imagery available.

Imaging spectrometry

An exciting development being planned by NASA is imaging spectrometry, which is the simultaneous acquisition of images in many narrow and contiguous spectral bands.

The existing satellites only produce a few broad, widely spaced spectral bands and this often results in the undersampling of the spectrum for each individual pixel. By increasing the number of spectral bands a near continuous spectrum could be created for each pixel. The analysis of these images will be greatly improved because of the wider selection of bands available, between 100 and 200 are planned, when compared with the present broadly selected general bands up to 7 in number.

The High Resolution Imaging Spectrometer (HIRIS) is proposed to be put into orbit in 1994 following a programme which started with airborne systems in 1983. HIRIS will operate in the 0.4 μm to 2.5 μm range with 128 bands covering a swath of 50 km and giving a 30 m resolution.

The National Remote Sensing Centre

As the UK interest in remote sensing expanded during the late 1970s it led to the formation of the National Remote Sensing Centre (NRSC) at the Royal Aircraft Establishment in Farnborough on 1 April 1980. The work of the centre is jointly funded by the following bodies:

Department of Trade and Industry
Overseas Development Administration
Department of the Environment
Ministry of Agriculture, Fisheries and Food
Scottish Development Department
Royal Society

The Centre is administered by the National Remote Sensing Programme Board whose membership consists of the contributing organizations, representatives of comercial users, the Natural Environment Research Council, the Science and Engineering Research Council, the Ministry of Defence, the Meteorological Office and the Remote Sensing Society. Control of the Centre at policy level is vested in this Board, which exercises financial control and approves the programme of work from year to year.

The manager of the Centre reports to the Programme Board and acts as Chairman of the Remote Sensing Applications Committee (RSAC) which gives him assistance and advice in the formulation of the Centre's programme. The membership of this committee comprises the Chairmen of the seven working groups which have been set up to foster co-operation between the NRSC and its users. These Working Groups are given in Figure 22.

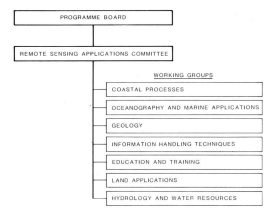

Figure 22. The organization of the UK National Remote Sensing Centre.

The Working Groups concern themselves with specific undertakings related to their titles and promote these activities with the support of the NRSC and the RSAC. The members of the Working Groups are drawn from government departments, academic institutes, and commercial and industrial organizations. The role of the NRSC can be summarized as follows:

1. To supply remote sensing data and imagery.
2. To provide facilities for research into and development of image processing, analysis and interpretative techniques.
3. To act as a focal point for the development of remote sensing techniques and their applications.
4. To provide education and training facilities in remote sensing.

To fulfil these objectives the NRSC is carrying out a campaign to promote user awareness and demonstrate potential applications of satellite imagery and thus maintains the following facilities to assist users:

● Accounts for the supply of imagery in either digital or photographic form from other remote sensing centres.
● An archive of imagery covering the UK and many other areas of the World. This data is stored in both digital and photographic form with over 1000 Landsat scenes being available.
● Browse file facilities to enable users to ascertain what imagery can be obtained.
● Film writers for converting digital data into photographic products.
● Photographic facilities for development and printing of film under precision quality control conditions.
● A microdensitometer for obtaining digital imagery from film.
● A general purpose digital image processing system providing a wide range of algorithms for customer use.
● A range of interactive image interpretation and display facilities.

The National Remote Sensing Centre also acts as the UK National Point of Contact (NPOC) with Earthnet, the organization within the European Space Agency responsible for the collection and dissemination of satellite remote sensing data in Europe, as shown in Figure 23. In addition, the NRSC will be a distribution centre for SPOT data and will maintain an archive of UK imagery. In preparation for this the NRSC organized a SPOT simulation campaign in 1984. The study covered approximately 40 sites in England, Wales, Scotland and Northern Ireland and involved many experienced remote sensing researchers and several first time users. This applications orientated campaign included studies of agriculture, forestry, urban and rural land use, vegetation, geology, hydrology, coastal processes, pollution and civil engineering.

To promote applications using remotely sensed data the NRSC performs many collaborative projects with users. Recent projects include hydrographic mapping in the Red Sea, pollution monitoring in the Arabian Gulf, woodland mapping in Southern England, land use and snow monitoring in Scotland and geological studies in various parts of Africa.

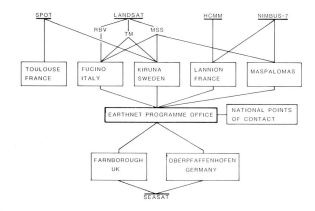

Figure 23. Satellite data reception and processing facilities in Europe.

For many applications it is necessary to relate imagery to maps. Such transformations require the selection of ground control points, which are features that can be readily identified on both the image and a map. The NRSC has formed a library of over 5000 Landsat MSS ground control points covering various European countries and has used the transformed imagery to make digitial mosaics of the UK, Ireland, Norway and Northern France.

The Ordnance Survey

The Ordnance Survey (O.S.) was founded in 1791 under the Board of Ordnance of the British Army for the purpose of producing military maps. The O.S. is now a civilian organization and is the National Mapping Agency for the United Kingdom responsible for the survey and mapping of the country, including geodetic and associated scientific work, topographical surveys and the publication of an extensive range of maps from these surveys.

From its purpose-built main office at Southampton together with over 150 local survey offices and a staff of approximately 3000 the O.S. undertakes its duty of mapping and revising the National Series together with specialist maps and its other specialist duties.

Maps are based on a framework of control points, both horizontal and vertical. The trigonometrical survey was started in the 19th century and retriangulation took place between 1935 and 1962. Primary triangulation provides a network of triangles with sides from 30 to 50 km in length but both shorter and longer sides occur because of the need to select prominent points on which to erect the pillars. The primary network is broken down into a secondary network with sides from 8 to 12 km in length which is further broken down to the tertiary network with sides from 4 to 7 km in length. In built-up areas additional control points are provided at intervals of 1.5 to 2 km for large scale mapping. The O.S. has nearly 22,000 control points and it has the responsibility to maintain these and to supply the horizontal position (co-ordinates) of these ground marks together with details of intervisibility with other points and other information. The values of the O.S. control points are based on a modified form of the Transverse Mercator or Gauss Conformal projection, and the maps are subsequently mapped on this projection. The Transverse Mercator projection suits a country like Britain where the greatest length is in the north/south direction. The central meridian, or line of longitude, which is selected as a line bisecting the country will have no scale error and distortion will be equal at the same distance east and west of this line. The Transverse Mercator projection can be thought of as a cylinder into which the Earth is fitted such that the line of longitude that is selected as the central meridian is in contact with the inside of the cylinder. The surface of the country is then projected onto the cylinder which is then cut open and up into sheets which become the maps.

To reduce the mapping error at the extremities of the country, that is east and west of the central meridian with the Transverse Mercator projection, a 'scale factor' is introduced which involves the reduction of dimensions on the spheroid (the Earth) by 0.04% before calculations are undertaken. The result of this reduction is to decrease the distances along the meridian and an enlargement by the same amount at the edge of the projection.

The vertical control is provided by a level network divided into geodetic and secondary levelling lines further broken down into bench marks to provide local vertical control. The level network is based on the Ordnance Datum which is the mean sea level of the sea at Newlyn in Cornwall.

With the horizontal and vertical control providing a network of positions on the surface of the Earth the detail in between can then be 'tied' into the national system.

The National Series which is provided by the O.S. is as follows:

1:1250	Town Maps
1:2500	Countryside Maps
1:10,000	Mountain and Moorland Maps
1:25,000	Pathfinder Maps
1:50,000	Landranger Maps
1:250,000	Routemaster Maps
1:625,000	Route Planning Maps
1:1,000,000	International Map of the World Series

The aim of the O.S. is to produce and maintain an up-to-date survey at 1:1250 for major urban areas and at 1:2500 and 1:10,000 scales for the remainder of the country. There are at present 53,315 town maps at 1:1250 scale, 157,987 at 1:2500 scale and 10,180 at 1:10,000 scale. The 1:1250 scale maps show houses, road names, house numbers and names, administrative boundaries, height control and archaeological information. The 1:2500 scale give details of parcel numbers and land areas. The 1:10,000 scale maps, which cover the whole country, show contours in colour at 10 m intervals in mountainous areas and 5 m interval elsewhere. The smaller scale maps are more well known and are more readily available from bookshops and newsagents; the O.S. has over 5000 outlets for its maps and other products. Other map series not mentioned above are the Outdoor Leisure maps covering nearly 30 of Britain's popular recreation areas; the Archaeological maps covering historical sites at various scales; the Town and City maps at 1:10,000 scale; and the Tourist maps at 1:50,000, 1:63,360 or 1:126,720 scale. In addition to the maps the O.S. provides Atlases and Guides.

Of interest to those using remote sensing data is the increasing availability of digital mapping at 1:1250, 1:2500 and 1:10,000 scale. The maps thus produced are therefore computer-based and rely on 'digital' techniques. It is at this stage that other information which is geographically referenced can be integrated with the available maps. The digital map can be produced at different scales and projections, with certain features selected in isolation to others dependent upon the needs of the user. A planner could show all the roads and railways in a region to assist in selecting alternative communications and the map produced would not be cluttered with extra information.

The O.S. has a representative on the Programme Board of the National Remote Sensing Centre and others are members of Working Groups of the NRSC.

Directory

This directory lists those academic institutions, government bodies and industrial and commercial companies in the United Kingdom which are associated with remote sensing.

Academic institutions

Department of Geography
University of Aberdeen
St Mary's
Old Aberdeen AB9 2UF

Remote Sensing Unit
Civil Engineering Department
Aston University
Gosta Green
Birmingham B4 7ET

Department of Geography
Bedford College
Regent's Park
London NW1 4NS

Department of Electronic & Electrical Engineering
University of Birmingham
PO Box 363
Birmingham B15 2TT

Geography Department
University of Birmingham
PO Box 363
Birmingham B15 2TT

Department of Geography
University of Bristol
University Road
Bristol BS8 1SS

Committee for Aerial Photography
University of Cambridge
Mond Building
Free School Lane
Cambridge

Department of Civil Engineering
The City University
Northampton Square
London EC1V 0HB

Department of Atmospheric Physics
Clarendon Laboratory
University of Oxford
Oxford OX1 3PU

Department of Electrical Engineering and Electronics
University of Dundee
Dundee DD1 4HN

Civil Engineering Department
University of Dundee
Dundee DD1 4HN

Carnegie Laboratory of Physics
University of Dundee
Dundee DD1 4HN

Department of Geography
University of Durham Science Laboratories
South Road
Durham DH1 3LE

School of Environmental Sciences
University of East Anglia
Norwich NR4 7TJ

Department of Geography
University of Edinburgh
Edinburgh EH1 1NR

Department of Geography
University of Exeter
Amory Building
Rennes Drive
Exeter EX4 4RJ

Department of Geography
Glasgow University
Glasgow G12 8QQ

School of Engineering
Hatfield Polytechnic
PO Box 109
Hatfield
Herts AL10 9AB

Physics Department
Heriot Watt University
Riccarton
Corrie
Edinburgh EH14 4AS

Computer Application Services
Heriot Watt University
79 Grassmarket
Edinburgh EH1 2HJ

Department of Applied Physics and Physics
Hull University
Hull HU6 7RX

Department of Geography
Hull University
Cottingham Road
Hull HU6 7RX

Centre for Remote Sensing
The Blackett Laboratory
Imperial College of Science and Technology
Prince Consort Road
London SW7 2BZ

Geology Department
Royal School of Mines
Imperial College of Science and Technology
Prince Consort Road
London SW7 2BP

Department of Earth Sciences
The Open University
Walton Hall
Milton Keynes MK7 6AA

Department of Surveying & Geodesy
University of Oxford
62 Banbury Road
Oxford OX2 6PN

School of Maritime Studies
The Polytechnic
Drake Circus
Plymouth PL4 8AA

Department of Geography
Portsmouth Polytechnic
Lion Terrace
Portsmouth PO1 3HE

Division of Surveying & Planning
Preston Polytechnic
Preston PR1 2TQ

Physics Department
Queen Elizabeth College
Campden Hill Road
London W8 7AH

Department of Geography
Queen's University
Belfast BT7 1NN

Department of Geography
University of Reading
Earley Gate
Whiteknights
Reading RG6 2AU

School of Mechanical & Offshore Engineering
Robert Gordon's Institute of Technology
Schoolhill
Aberdeen AB9 1FR

Department of Military Technology
Royal Military Academy Sandhurst
Faraday Hall
Camberley
Surrey GU15 4PQ

Civil Engineering Department
Royal Military College of Science
Shrivenham
Swindon
Wilts SN6 8LA

Department of Geology
The University of St Andrews
Fife

Department of Geography/Geology
College of St Mark and St John
Derriford Road
Plymouth PL6 8BH

Department of Geography
School of Oriental & African Studies
Malet Street
London WC1E 7HP

Department of Geography
University of Sheffield
Western Bank
Sheffield S10 2TN

Department of Geography
University of Southampton
Southampton SO9 5NH

Department of Oceanography
University of Southampton
University Road
Southampton SO9 5NH

Department of Civil Engineering
University of Surrey
Guildford GU2 5XH

Department of Geography
University College London
Gower Street
London WC1 6BT

Department of Photogrammetry & Surveying
University College London
Gower Street
London WC1E 6BT

Department of Construction
Oxford Polytechnic
Headington
Oxford OX3 0BP

Department of Physics & Astronomy
University College London
Gower Street
London WC1E 6BT

Department of Geography
Llandinan Building
University College of Wales
Penglais
Aberystwyth
Dyfed SY23 3DB

Department of Physical Oceanography
University College of North Wales
Marine Science Laboratories
Menai Bridge
Gwynedd

Department of Geography
University College Swansea
Singleton Park
Swansea SA2 8PP

Department of Oceanography
University College Swansea
Swansea SA2 8PP

Government bodies

Ministry of Agriculture, Fisheries & Food
Agricultural Development Advisory Service
Aerial Photography Unit
Block B
Brooklands Avenue
Cambridge CB2 2DR

Hydrographic Department
Ministry of Defence
Taunton TA1 2DN

Department of Energy
Thames House South
Millbank
London SW1P 4QJ

Planning Policy, Minerals and New Towns
Department of the Environment
Prince Consort House
27/29 Albert Embankment
London SE1 7TF

Space Branch
Department of Trade and Industry
29 Bressenden Place
London SW1E 5DT

Air Photographs Unit
Scottish Development Department
Graphics Group
New St Andrew's House
St James Centre
Edinburgh EH1 3SZ

Air Photographs Unit
Welsh Office
Room G-003
Crown Offices
Cathays Park
Cardiff CF1 3NQ

British Antarctic Survey
Madingley Road
Cambridge CB3 0ET

Central Electricity Research Laboratories
Central Electricity Generating Board
Kelvin Avenue
Leatherhead
Surrey KT22 7SE

Overseas Development Administration
Kingston Road
Tolworth
Surrey KT5 9NS

Exmoor National Park
Exmoor House
Dulverton
Somerset TA22 9HL

Research and Development Division
Forestry Commission
Field Survey Branch
Alice Holt Lodge
Wrecclesham
Farnham
Surrey

Grampian Regional Council
Woodhill House
Ashgrove Road West
Aberdeen

Image Analysis and Remote Sensing Group
Building 521
Harwell Laboratory
Didcot
Oxon OX11 0RA

Department of Pure and Applied Biology
Imperial College of Science and Technology
Prince Consort Road
London SW7 2BB

Department of Geography
University of Keele
Keele
Staffs ST5 5BG

Department of Geography
University of London
King's College
Strand
London WC2

Department of Geography
Kingston Polytechnic
Penrhyn Road
Kingston upon Thames KT1 2EE

Department of Environmental Sciences
University of Lancaster
Lancaster LA1 4YQ

Department of Earth Sciences
The University
Leeds LS2 9JT

School of Geography
The University
Leeds LS2 9JT

Department of Astronomy
University of Leicester
University Road
Leicester LE1 7RH

Department of Maritime Studies
Liverpool Polytechnic
Byrom Street
Liverpool L3 3AF

Department of Geography
University of Liverpool
Roxby Building
PO Box 147
Liverpool L69 3BX

Department of Geography
London School of Economics and Political Science
Houghton Street
London WC2A 2AE

Department of Civil & Structural Engineering
Institute of Science & Technology
University of Manchester
PO Box 88
Sackville Street
Manchester M60 1QD

Department of Geography
University of Manchester
Manchester M13 9PL

National College of Agricultural Engineering
Silsoe
Beds MK45 4DT

Department of Geology
University of Newcastle upon Tyne
Newcastle upon Tyne NE1 7RV

Department of Land Surveying
North East London Polytechnic
Longbridge Road
Dagenham
Essex RM8 2AS

Department of Geography & Geology
Polytechnic of North London
Marlborough Building
383 Holloway Road
London N7 0RN

Hydraulics Research Station
Howbery Park
Crowmarsh
Wallingford
Oxon OX10 8BA

Institute of Oceanographic Sciences
Wormley
Godalming
Surrey

Remote Sensing Unit
The Macaulay Institute for Soil Research
Craigiebuckler
Aberdeen AB9 2QJ

Directorate of Military Survey
Survey 4
Elmwood Avenue
Feltham
Middx

NERC Scientific Services
Holbrook House
Station Road
Swindon SN1 1DE

Photogeology Unit
NERC Institute of Geological Sciences
Overseas Division
Keyworth
Nottingham NG12 5GG

Meteorological Office Radar Research Laboratory
Royal Signals and Radar Establishment
St Andrews Road
Great Malvern
Worcs

National Maritime Institute
Faggs Road
Feltham
Middx TW14 0LQ

NRSC Space Department
Q134 Building
RAE
Farnborough
Hants GU14 8TD

Ordnance Survey
Romsey Road
Maybush
Southampton SO9 4DM

Royal Commission on Historical Monuments
National Monuments Record
Air Photographs Unit
Fortress House
23 Savile Row
London W1X 1AB

Rutherford and Appleton Laboratories
Chilton
Didcot
Oxon OX11 0GX

Soil Survey Of England and Wales
Photo Interpretation Units
Rothamsted Experimental Station
Harpenden
Herts AL5 2JQ

Transport and Road Research Laboratory
Crowthorne
Berks RG11 6AU

Oil Pollution Division
Warren Spring Laboratory
Gunnels Wood Road
Stevenage
Herts SG1 2BX

Industry and commerce

Aeronautical & General Instruments Ltd
40 Purley Way
Croydon CR9 3BH

AGA Infrared Systems
Arden House
West Street
Leighton Buzzard
Beds LU7 7ND

Analysis Automation Ltd
Southfield
Eynsham
Oxford OX8 1JD

Barr & Stroud
Caxton Street
Anniesland
Glasgow G13 1HZ

Bell & Howell Ltd
Electronics & Instruments Division
Lennox Road
Basingstoke
Hants RG22 4AW

B K S Surveys Ltd
47 Ballycairn Road
Coleraine
Co Londonderry BT51 SHZ

BMMK & Partners
Herontye
East Grinstead
Sussex RH19 4QA

Bristol Industrial & Research Associates Ltd
PO Box 2
Portishead
Bristol BS20 9JB

British Aerospace Plc
Gunnels Wood Road
Stevenage
Herts SG1 2AS

British Air Survey Association
c/o J. A. Story & Partners
92-94 Church Road
Mitcham
Surrey CR4 3TD

The British Petroleum Company
Moor Lane
London EC2Y 9BU

Brown's Geological Information Service Ltd
160 North Gower Street
London NW1 2ND

Carl Zeiss Jena
C Z Scientific Instruments Ltd
PO Box 43
Elstree Way
Borehamwood
Herts WD6 1NH

Cartographic Engineering Ltd
Landford Manor
Salisbury
Wilts SP5 2EW

Cartographic Services (Southampton) Ltd
Landford Manor
Salisbury
Wilts SP5 2EW

Cartosat Production Ltd
Lansing Building
Old Station Yard
Marlpit Hill
Edenbridge
Kent TN8 5AU

Technical Department
Charter Consolidated Ltd
40 Holborn Viaduct
London EC1

Clyde Surveys
Reform Road
Maidenhead
Berks SL6 8BU

Computer Application Services
Heriot-Watt University
79 Grassmarket
Edinburgh EH1 2HJ

Consolidated Goldfields
40 Moorgate
London EC2

CW Controls Ltd
Industrial Electronics Engineers
Jubilee Works
Crowland Street
Southport
Merseyside PR9 7RR

Earth Data Limited
Unit 19
South Hampshire Industrial Park
Salisbury Road
Totton
Southampton SO4 3SA

EASAMS Limited
Lyon Way
Frimley Road
Frimley
Surrey

The Economist Intelligence Unit Ltd
Spencer House
27 St James's Place
London SW1A 1NT

EMI Electronics Ltd
Victoria Road
Feltham
Middx

ERSAC Ltd
Peel House
Ladywell
Livingston
W. Lothian EH54 6AG

Esso Research Centre
Abingdon
Oxon OX13 6AE

Feedback Instruments Ltd
Park Road
Crowborough
Sussex TN6 2QR

Fisher Spence Associates
4 Ardross Street
Inverness IV3 5NN

Focal Point Audiovisual Ltd
251 Copner Road
Portsmouth PO3 5EE

GEMS of Cambridge Ltd
CAD Centre
Madingley Road
Cambridge CB3 0HB

General Technology Systems Ltd
Forge House
20 Market Place
Brentford
Middx TW8 8EQ

Geosurvey International Ltd
Geosurvey House
Orchard Lane
East Molesey
Surrey KT8 0BY

Gresham Lion (PPL) Ltd
Lower Way
Thatcham
Berks RG13 4RE

Hunting Geology and Geophysics
Elstree Way
Borehamwood
Herts WD6 1SB

Hunting Surveys Ltd
Elstree Way
Borehamwood
Herts WD6 1SB

Hunting Technical Services
Elstree Way
Borehamwood
Herts WD6 1SB

The IAL Group
Aeradio House
Hayes Road
Southall
Middx UB2 5NJ

Intergraph (Great Britain) Ltd
Albion House
Oxford Street
Newbury
Berks RG13 1JG

International Instrumentation
Marketing Co Ltd
98 High Street
Thame
Oxon OX9 3EH

Intertrade Scientific Ltd
Mill House
Boundary Road
Loudwater
High Wycombe
Bucks

Joyce-Loebl
Marquisway
Team Valley Estate
Gateshead
Tyne and Wear NE11 0QW

Kodak Limited
PO Box 66
Hemel Hempstead
Herts HP1 1JU

Linotype-Paul Ltd
Scanner Division
Sandford Park Trading Estate
Corpus Street
Cheltenham
Glos GL52 6XH

Logica Limited
Cobham Park
Cobham
Surrey KT11 3LX

McCarta Ltd
122 Kings Cross Road
London WC1X 9DS

McMichael Ltd
Environmental Data Systems Group
Wexham Road
Slough SL2 5EL

Map Services
The Old Post Office
The Street
Terling
Essex CM3 2PG

Marconi Research Laboratories
West Hanningfield Road
Great Baddow
Chelmsford CM2 8HN

Marconi Space & Defence Systems Ltd
Browns Lane
The Airport
Portsmouth PO3 5PH

Masdar (UK) Ltd
141 Nine Mile Ride
Finchampstead
Wokingham
Berks RG11 4HY

Meridian Airmaps Ltd
Marlborough Road
Lancing
Sussex BN15 8TT

Microdata Ltd
Monitor House
Station Road
Radlett
Herts WD7 8JX

Nigel Press Associates Ltd
Apex House
High Street
Edenbridge
Kent TN8 5AU

NMI Ltd
Faggs Road
Feltham
Middx

Dr J. W. Norman
35 Brangwyn Avenue
Brighton BN1 8XH

Pilkington-PE Ltd
Glascoed Road
St Asaph
Clwyd LL17 0LL

Programmed Neuro Cybernetics (UK) Ltd
Clivia House
65 Old Church Street
London SW3 5BS

Racal Recorders Ltd
Hardley Industrial Estate
Hythe
Southampton SO4 6ZH

Rae-Thomson & Co
17 Stanley Street
Aberdeen AB1 6US

Remote Sensing Products and Publications Ltd
54 Hamilton Street
Broughty Ferry
Dundee

Riofinex Ltd
6 St James's Square
London SW1

Rio Tinto Finance & Exploration Ltd
PO Box 133
6 St James's Square
London SW1Y 4LD

Robertson Research International Ltd
Ty'n-y-Coed
Llanrhos
Llandudno
LL30 1SA

Rofe, Kennard & Lapworth
Consulting Engineers
Raffety House
2/4 Sutton Court Road
Sutton
Surrey SM1 4SS

Ross Instruments Ltd
Morgans Vale Road
Redlynch
Salisbury
Wilts SP5 2HU

Selection Trust Ltd
Lower William Street
Northam
Southampton SO1 1QE

Sigmex Ltd
Sigma House
North Heath Lane
Horsham
W. Sussex RH12 4UZ

Smith Associates
Consulting System Engineers Ltd
45-47 High Street
Cobham
Surrey KT11 3DP

Soil Mechanics Ltd
Foundation House
Eastern Road
Bracknell
Berks RG12 2UZ

Space Frontiers Ltd
30 Fifth Avenue
Havant
Hants PO9 2PL

Spectrascan Ltd
31a North Street
Emsworth
Hants PO10 7DA

Spembly Ltd
5 Vicarage Hill
Alton
Hants GU34 1HT

SRS Consultants
40 Maisemore Gardens
Emsworth
Hants PO10 7JX

J. A. Story & Partners
92-94 Church Road
Mitcham
Surrey CR4 3TD

Survey & General Instruments
Fircroft Way
Edenbridge
Kent

Systems Designers Ltd
Systems House
1 Pembroke Buildings
Camberley
Surrey GU13 8NZ

Technical Contracting Services
34 Third Avenue
Hove
E. Sussex BN3 2EB

Videlcom Ltd
33 Bruton Street
London W1X 7DD

W. Vinten Ltd
Western Way
Bury St Edmunds
Suffolk IP33 3TB

Wild Heerbrugg (UK) Ltd
Revenge Road
Lordswood
Chatham
Kent ME5 8TE

Wimpey Laboratories Ltd
Beaconsfield Road
Hayes
Middx UB4 0LS

Glossary

Active remote sensing	Remote sensing techniques which provide their own source of electromagnetic energy, e.g., radar
Atmosphere	The layers of gases that surround the Earth
Atmospheric window	Wavelength inteval which is not absorbed by gases in the atmosphere
Colour-composite image	A colour image produced by projecting different colours through each different waveband to produce a 4-colour image when registered
Density of image	A measure of opacity or brightness of an image, i.e., tonal value
Emittance	The radiation of electromagnetic energy from a material
Geostationary	Rotating with the Earth, having an apparently fixed position
Grey scale	A calibrated sequence of grey tones ranging from white to black. Landsat MSS imagery has 256 different tones, each of which is given an ordered number according to its tonal value or density
Ground swath	The area on the ground which is surveyed by the scanner system
Hardware	Instruments used in remote sensing image analysis: equipment, scanners, etc.
Image	The representation of a scene as recorded by a remote sensing system
Instantaneous field of view	The resolution of a scanning system
Interpretation	Information extraction from an image
Infrared	Part of the electromagnetic spectrum, 0.7–300 mm
Microwave	Part of the electromagnetic spectrum, 0.3–300 cm
Multispectral (also multiband)	the use of more than one band in a remote sensing system
Nadir	The point vertically below an object, e.g., the point on the Earth's surface below a satellite
Orthophoto	Photograph which has been rectified differentially, i.e., corrected for relief and tilt displacement, to produce a uniform scale
Panchromatic film	Film sensitive to visible light
Passive remote sensing	Remote sensing of energy naturally reflected or radiated from the terrain
Photogrammetry	The art and science of taking measurements from photographs and relating them to the 'real' world or to maps
Photomap	An aerial photograph which has been enhanced sufficiently to be used as a topographic map i.e., a map with a photographic background

Pixel	A single picture element
Radar	Radio detection and ranging: an active form of remote sensing utilizing the 0.1–100 cm part of the spectrum
Radiation	The propagation of energy in the form of electromagnetic waves
Real time	Simultaneous inspection of data at time of recording
Reflectance	The amount of energy reflected compared to the amount absorbed
Remote sensing	The collection, processing and interpretation of data related to the Earth which is detected without physical contact; technically, it is restricted to some parts of the electromagnetic spectrum
Resolution	The ability to distinguish between objects on the image
Réseau	The intersection of grid lines contained on RBV images and some aerial photographs
Scanner	An instrument used to move the detector so that a whole area is observed as the aircraft flies over; it builds an image, pixel by pixel, line by line
Software	Computer programs
Spectrum	The electromagnetic spectrum: a continuous sequence of energy arranged according to wavelength or frequency
Swathwidth	The surface width covered by a scanner

Abbreviations

ATM	Airborne thematic mapper
AVCS	Advanced vidicon camera system
AVHRR	Advanced very high resolution radiometer
CCT	Computer compatible tape
DCS	Data collection system (on Landsat)
EARSeL	European Association of Remote Sensing Laboratories
EEC	European Economic Community (Common Market)
EMS	Electromagnetic spectrum
EREP	Earth resources experiment package
EROS	Earth resources observation system
ERTS	Earth resources technology satellite (renamed Landsat)
ESSA	Environmental Science Services Administration
GDTA	Groupement pour le Développement de la Télédétection
IR	Infrared
MSS	Multi-spectral scanner
NASA	National Aeronautical and Space Administration
NOAA	National Oceanic and Atmospheric Administration
NPOC	National Point of Contact
NRSC	National Remote Sensing Centre (Farnborough, UK)
RADAR	Radio detection and ranging
RBV	Return beam vidicom (Landsat 1-3)
SAR	Synthetic aperture radar
SLAR	Side looking airborne radar
SONAR	Sound navigation and ranging
SPOT	Systeme probatoire d'observation de la terre
TIR	Thermal infrared
TIROS	Television and infrared observation satellite
TM	Thematic mapper (on Landsat 4 and 5)
UV	Ultraviolet
VDU	Visual display unit
WBVTR	Wide band video tape recorder

Index